MW01517905

Constructed Wetland for Domestic Wastewater Treatment

Birhanu Genet
Dr. Seyoum Leta

Constructed Wetland for Domestic Wastewater Treatment

A Case Study in Addis Ababa, Ethiopia

LAP LAMBERT Academic Publishing

Impressum/Imprint (nur für Deutschland/only for Germany)
Bibliografische Information der Deutschen Nationalbibliothek: Die Deutsche Nationalbibliothek verzeichnet diese Publikation in der Deutschen Nationalbibliografie; detaillierte bibliografische Daten sind im Internet über http://dnb.d-nb.de abrufbar.

Coverbild: www.ingimage.com

Verlag: LAP LAMBERT Academic Publishing GmbH & Co. KG
Dudweiler Landstr. 99, 66123 Saarbrücken, Deutschland
Telefon +49 681 3720-310, Telefax +49 681 3720-3109
Email: info@lap-publishing.com

Herstellung in Deutschland:
Schaltungsdienst Lange o.H.G., Berlin
Books on Demand GmbH, Norderstedt
Reha GmbH, Saarbrücken
Amazon Distribution GmbH, Leipzig
ISBN: 978-3-8465-0439-0

Imprint (only for USA, GB)
Bibliographic information published by the Deutsche Nationalbibliothek: The Deutsche Nationalbibliothek lists this publication in the Deutsche Nationalbibliografie; detailed bibliographic data are available in the Internet at http://dnb.d-nb.de.

Cover image: www.ingimage.com

Publisher: LAP LAMBERT Academic Publishing GmbH & Co. KG
Dudweiler Landstr. 99, 66123 Saarbrücken, Germany
Phone +49 681 3720-310, Fax +49 681 3720-3109
Email: info@lap-publishing.com

Printed in the U.S.A.
Printed in the U.K. by (see last page)
ISBN: 978-3-8465-0439-0

Addis Ababa University

School of Graduate Studies

Environmental Science Program

Constructed Wetland Systems for Domestic Wastewater Treatment: A Case Study in Addis Ababa, Ethiopia

A thesis submitted to the School of Graduate Studies, Addis Ababa University

In Partial Fulfillment of the Requirements for the Degree of Masters of Science in Environmental Science

By: Birhanu Genet Fenta

Table of Contents

LIST OF TABLES

LIST OF FIGURES

ACRONYMS

AAWSA	Addis Ababa Water Supply and Sewerage Authority
APHA	American Public Health Association
BOD_5	Biochemical Oxygen Demand
CFU	Colony Forming Unit
COD	Chemical Oxygen Demand
CBO	Community Based Organizations
CW	Constructed Wetland
CPC	Cleaner Production Center
DO	Dissolved Oxygen
EEPA	Ethiopian Environmental Protection Authority
EIBC	Ethiopian Institute of Biodiversity Conservation.
ESTA	Ethiopian Science and Technology Agency
FC	Fecal Coliforms
GTZ	German Agency for Technical Cooperation
SF	Surface Flow
SSF	Subsurface Flow
HRT	Hydraulic Retention Time
IUCN	International Union for Conservation of Nature and Natural Resources
JWBO	Jehovah's Witnesses Branch Office
L	Litter
MF	Membrane Filter
Ml	Milliliter
Mg	Milligram
TC	Total Coli forms
TN	Total Nitrogen
TP	Total Phosphorous
TSS	Total Suspended Solids
USEPA	Unite Sates of America Environmental Protection Agency
UNDP	United Nations Development Program
UNIDO	United Nations Industrial Development Organization

ACKNOWLEDGEMENTS

First and foremost, I would like to express my gratitude to my advisor, Dr. Seyoum Leta for his continuous guidance, invaluable suggestions and encouragement throughout the study. Without his demonstrative supervisions at every stage, this work could never have been materialized.

I would also like to thank Mr. Afework Hailu, the Director of Ethio-Wetland and Natural Resources Association and Mr. Mohammed Ali, Head, Pollution Control Department of FEPA of Ethiopia, who have guided my effort through contribution of conceptual ideas and comments by intensively reading the document starting from project proposal to the final document of the study.

Sincere thanks to Horn of Africa Regional Environmental Center/Network for financial support with its Demand Driven Action Research Program that enabled me to perform the hard tasks related with this research works in a sound way. In addition to this, I would like to thank Jehovah's Witnesses Branch Office leaders & workers who gave me permission to do this study on their CW and allowed me to take different interested persons such as Ethio-Wetland and Natural Resource Association and Addis Ababa Water supply and Sewerage Authority staff members to visit the site.

I also extend my deep appreciation to all staff members of Applied Microbiology Laboratory of Biology Department, Addis Ababa University and Federal Environmental Protection Authority as well as Addis Ababa Environmental Protection Authority Laboratory Staffs for their kind cooperation and permission to use their laboratory services for sample analyses.

Last but for not least, I want to thank GOD for helping me to complete my thesis in due time. Thank you for making everything possible and showing grace and your love in times when I needed it the most. Next to GOD, Very special thanks to my family and all friends, who have supported and encouraged me through all phases of my life

ABSTRACT

During the last decades, constructed wetlands were very successful when used for treatment of wastewater from different sources such as municipal, domestic, industrial, agricultural and surface runoff. This new approach is designed based on natural processes involving complex and concerted interactions between the plants, the substrate and the inherent microbial community to accomplish wastewater treatment in a more controlled and predictable manner through physical, chemical and biological processes.

While CW have a proven effectiveness for treatment of a variety of wastewaters in developed countries, little work has been done in developing countries including Ethiopia where the concept of constructed wetlands for wastewater treatment is still a relatively new idea. Therefore, the main objective of this study was to evaluate the treatment performance of constructed wetland as an alternative municipal wastewater treatment technology under Ethiopian climatic conditions, by taking JWBO CW as a case study.

To verify the efficiency of JWBO CW, a total of 24 samples were collected and analyzed for selected wastewater quality parameters. Biochemical oxygen demand, chemical oxygen demand, total suspended solids, ammonium N, nitrate N, total N, orthophosphate, total phosphorus, sulfate, sulfide, total coliform and fecal coliform were all measured using standard methods. Wastewater temperature and pH was measured on-site.

The treatment performance of JWBO wetland was evaluated based on the percentage removal of selected wastewater quality parameters. Within the study period, the mean removal efficiency of JWBO CW system was 99.3% (BOD_5), 89% (COD), 85% (TSS), 28.1% (NH_4^+-N), 64% (NO_3-N), 61.5% (TN), 28% (orthophosphate), 22.7% (TP), 77.3% (Sulfate), 99% (Sulfide), 94.5% (TC) and 93.1% (FC).

Moreover, the result of this study indicated that wetland cells planted with *Cyprus papyrus* (cell 1 and 3) showed higher removal efficiency for NO_3-N (82.4%), NH_4^+-N (24.8%), TN (54.8%), PO_4^{3-} (23.5%), and TSS (83.9%) than the other wetland cells. Similarly wetland cells planted with *Phoenix canariensis* (cell 4 and 6) showed higher removal efficiency for TP (17%), S^{2-} (99%), BOD_5 (98%), COD (90%), TC (94%) and FC (91%). While the other

viii

wetland cells planted with *Cyprus alternifolia* (cell 2 and 5) showed higher removal efficiency only for SO_4^{2-} (82.2%) than the others.

The performance efficiency results indicated that, this wetland system has excellent removal capability for biochemical oxygen demand, chemical oxygen demand, total suspended solids, sulfate, sulfide, total and fecal coliform bacteria. However, since the HRT of JWBO CW was very short (2.16 days) the removal efficiency was low for nitrogen (especially ammonium nitrogen) and phosphorus. Thus, for effective wastewater treatment performance, constructed wetlands should consist of a minimum of two to three cells in serious and all the cells should be vegetated with different plant species within the system.

In general based on the overall results of the treatment performance of JWBO CW, the application of constructed wetland in Ethiopia can be considered technically as well as economically viable option for domestic wastewater treatment.

1. INTRODUCTION

Background

The Ramsar Convention,(Iran, 1971; Article 1.1), defined wetland as "areas of marsh, fen, peatland, or water, whether natural or artificial, permanent or temporary, with water that is static or flowing, fresh, brackish or salt, including areas of marine water the depth of which at low tide doesn't exceed six meters." In addition, the convention (Article 2.1) provides that wetland may incorporate riparian and coastal zones adjacent to the wetlands and islands or bodies of marine water deeper than six meters at low tide lying within the wetland (Ramsar Convention Bureau, 1997).

Wetlands have been referred to as "Kidneys" of our environment (Wallance, 1998), "living machines" (MacDoland, 1994) and "...one of nature's most effective ways of cleansing polluted water" (Rocky mountain institute, 1998). They have been termed "Kidneys of the planet" because of the natural filtration processes that occur as water passes through.

All wetlands, natural or constructed, fresh-water or salt, have many distinguishing features, the most notable of which are the presence of standing water, unique wetland soils and plants adapted to or tolerant of saturated soils (William and James,1993; USEPA, 1993). In Ethiopian context marsh areas, swamplands, flood plains, natural and artificial ponds, volcanic creator lakes and upland bogs are treated collectively as wetland ecosystem (EIBC, 2007; Abebe and Geheb, 2004).

According to Luise *et al.* (1999), wetlands provide a number of functions and values; (Wetland functions are the inherent processes occurring in wetlands; wetland values are the attributes of wetland that the society perceives as beneficial). Under appropriate circumstances wetlands can provide; water quality improvement (William, 1997), cycling of nutrients (Nichils, 1983), habitat for fish and wildlife, flood storage and the resynchronization of storm rainfall and surface runoff (Ramsar Convention Bureau, 1997), Passive recreation such as bird watching and photography, active recreation (such as hunting, education and research) and aesthetics as well as landscape enhancement (Tanner and Sukias, 2003).

1

The recognition of these wetland values and the presence of policies such as "no net loss" of wetlands in some countries have stimulated construction of wetland that have specific objectives such as the mitigation of unavoidable wetland losses, wildlife enhancement, domestic wastewater treatment, mine drainage control, and storm water retention and control (William and James, 1993; Martha, 2003). Because of this fact, currently constructed wetlands are being used at increasing rate for treatment of wastewaters in different sources because of their consistent performance for pollutant removal (Renee, 2001; Muhammad *et al.*, 2004)

A "constructed wetland" is defined as human made, engineered areas specifically designed for the purpose of treating wastewater or storm water by establishing optimal physical, chemical and biological conditions that occur in natural wetland ecosystems (Hammer, 1989; USEPA, 1993; Luise *et al.*, 1993).

Constructed wetlands as an alternative wastewater treatment method

Studies of the feasibility of using constructed wetlands for wastewater treatment were initiated during the early 1950s in Germany, with the first operational horizontal subsurface flow constructed wetland appearing in 1974. In United States, wastewater treatment using either natural or constructed wetland researches began in the late 1960s and increased dramatically in the scope during 1970s (Joseph, 2005.)

During the last decades, constructed wetlands were very successful when used for wastewater, and low quality water treatment from different sources (Nicols, 1983; Chris and Vivian, 1997; Rechared, 1998)). This new approach is designed based on natural processes involving complex and concerted interactions between the plants, the substrate/media and the inherent microbial community to accomplish wastewater treatment in a more controlled and predictable manner through physical, chemical and biological processes (Simi and Mitchell, 1999; Joseph, 2005)

Because they are natural systems, CW are effective, reliable, simple, environmental friendly and relatively inexpensive to install and maintain (Gersberg *et al.*, 1984; Rogers *et al.,* 1991; Dewardar and Bahgat, 1995; Vymazai, 1996; Zuidervaart *,et al.,*1999; Coleman *et al.,* 2001; Vymazai, 2002.) They have been successfully applied worldwide for biological treatment of municipal and industrial wastewater (USEPA, 1988; Okurut *et al.,* 1999; Nzengya and Witshitemi, 2001; Kyambadde *et al.,* 2004), and agricultural wastewater (Kenneth, 2001) as well as surface runoff water (William, 1997).

For example in Uganda, the economic viability of using constructed wetland was deduced from the total annual cost of the wetland and waste stabilization ponds designed for a population equivalent of 4000. The result of this study indicated that the total annual cost for waste stabilization pond was 21% more than that of the constructed wetlands ($11,400) (TomOkin, 2000). The other study done in Ireland by Reddy (2004) showed that the cost of atypical constructed wetland with a size $4650m^2$ is about $122,000, which was cheaper by 30% than conventional treatment methods considering the lifespan (which is 30–50 years) and replacement values of the wetland.

The above case studies also confirmed that maintenance cost for constructed wetland was eight times lower than the conventional treatment systems. Based on the overall results of the treatment performance and costs, these researchers concluded that the application of constructed wetlands can be considered both technically and economically viable option for municipal wastewater treatment.

Additionally, CW attracts wildlife such as birds, mammals, amphibians, and variety of dragonflies and other insects make the wetland home (Martha, 2003). For instance, the recent USEPA publications (1999) indicated that more than 1,400 species of wildlife have been identified from constructed and natural treatment wetlands, of these more than 800 species were reported in CW alone. Moreover, constructed wetland plants, (especially when they are planted with ornamental plant species), provides a more aesthetically pleasing alternative than many other conventional wastewater treatment systems (Richard *et al.,* 1994).

3

Due to these benefits, over the past twenty years constructed wetlands have been used effectively to decrease the concentrations of various pollutants from different sources particularly in Europe and North America. Despite the numerous articles published on wetland in these countries, in recent years there is a notable gap in the literature regarding constructed wetlands in developing countries for wastewater treatment (Faithful, 1996)

For instance, within this ten years only limited wetlands for the treatment of wastewater have been constructed and studied in East Africa e.g. in Uganda, for treating municipal wastewater (Okurut *et al.*, 1999), in Tanzania, for treating wastewater from the waste stabilization ponds at the University of Dares Salaam (Mashauri *et al.*, 2000; Kaseva, 2003) and in Kenya, for domestic wastewater treatment (Oketch, 2000; Nyakango and VanBruggen, 2001). These studies have shown that constructed wetlands are very suitable for treatment of wastewater in tropical climates.

While constructed wetlands have such a proven effectiveness for treatment of a variety of wastewaters (Hester and Harrison, 1995; Joseph, 2005; Muhammad *et al.*, 2004), no work has been done in Ethiopia where the concept of constructed wetlands for wastewater treatment is still a relatively new idea. For instance, in Addis Ababa, the current wastewater treatment system (stabilization pond and sewer line) serves only a small part (2%) of the population (Getahun *et al.*, 1999; AAWSA, 2003) were arranged 60 years ago with the design capacity of 70,000 people, and the developments achieved since then are small (AAWSA, 2001; ESTA /CPC, 2004).

Due to this, approximately 73% of the inhabitants "disposed" feces and dirty waters in pit latrine or septic tank and a sizeable part of the population (25%) has no such facilities at all (AAWSA, 2003). In addition to these, more than half of the country's industries are found in Addis Ababa, but very few of them have a treatment plant or a connection to a sewer. These parts of the population and industries dispose their wastewater to natural watercourses, natural wetlands or elsewhere appearing them suitable (Getahun *et al.*, 1999; AAWSA, 2003; Afework, 2003).

To solve this problem the municipality in collaboration with GTZ (2004) developed five years strategic plan, (2004– 2008), for wastewater management options especially for the central part of the city. These options were divided in to two categories. The first one was centralized option, which was planned to expand the existing wastewater treatment plant, by improving the sewer system (Waste Stabilization Pond System) at Kality. The second option was to develop decentralized wastewater treatment options such as trickling filter, biogas digester, dry toilets with urine diversion and vermin composting tanks (AAWSA, 20003). But all of these conventional options, (both centralized and decentralized), would require high initial and operational costs, as well as needs skilled manpower for operation and maintenance (USEPA, 1993; Simi and Mitchell, 1999; Tanner and Sukias, 2003)

For developing countries like Ethiopia that have limited resources for the construction and operation of conventional treatment plants, there should be an option which is economical, but produce an effluent with same, even better quality from the conventional treatment system. Then the need for improvement and conservation of the environment in Ethiopia is necessitating the provision of energy and cost effective secondary wastewater treatment facilities for small communities such as schools, hospitals, military camps, colleges, farms, industries, and universities where on-site wastewater disposal technology is predominant.

Therefore, the primary purpose of this study was to demonstrate constructed wetlands as an alternative cost effective and environmentally friendly technology for domestic wastewater treatment for regulatory bodies, investors, academia, business institutions and other concerned bodies

Types of constructed wetlands and their treatment mechanisms
There are two main types of constructed wetlands: Surface flow (SF) constructed wetlands and Subsurface flow (SSF) constructed wetlands (Hammer, 1992; USEPA, 1993; Tchobanoglous, 1997).

1.3.1 Surface Flow Wetland

Surface flow constructed wetland systems most resemble natural wetlands both in the way they look and the way they provide treatment. Both designs can be used to treat wastewater from individual and community sources, but surface flow wetlands are usually more economical for treating large volumes of wastewater (Sinclair, 2000). The surface flow (SF) wetland typically consists of a shallow basin, soil or other medium to support the roots of plants and a water control structure that maintains a shallow depth of water (Luise *et al.*, 1999) (Figure 1)

As soon as wastewater enters to surface flow wetland cell, natural processes immediately begin to break down and remove the waste materials in the water (Renee, 2001; Kaseva, 2003). Before the wastewater has moved very far in the wetland small suspended waste materials are physically strained out by submerged plants, plant stems, and plant litter in the wetland (Hammer, 1992).

The roots, stems, leaves, and litter of wetland plants also provide a multitude of small surfaces where wastes can become trapped and waste-consuming bacterial can attach themselves to the plant (USEPA, 1993; Sinclair, 2000). These bacteria provide the majority of wastewater treatment. Wind, rain, wastewater and anything else that agitates the water surface can add oxygen to the system. This helps the aerobic bacteria thrive in wetlands near the surface wherever oxygen is present, in addition to this, anaerobic bacteria thrive in the wetland where there is little or no oxygen (USEPA, 1998)

When bacteria consume waste particles in the water they convert them into other substances such as methane, ammoniu, sulfate, orthophosphate, carbon dioxide and new cellular material. Some of these substances are used as food by plants and other by bacteria (Christina, 2005). For any of the processes in wetlands to work, the wastewater must remain in the system long enough for treatment to occur naturally. The hydraulic residence time for wastewater in SF systems is based on wastewater strength, the level of desired treatment and climatic factors

Plants help treatment processes of SF wetlands in several ways; filter wastes, regulate flow and provide surface area for bacterial treatment. Floating plants, such as water lilies and

6

emergent plants, such as cattails, shade the water surface and control algae growth (Sinclair, 2000). The advantages of SF wetlands over SSF wetlands are their construction, operation, and maintenance are straightforward. The main disadvantage of SF is its requirement of a larger land area than other systems (Luise *et al.*, 1999).

General characteristic of SF wetland:-
Water level is above the ground surface; vegetation is rooted and emerges above the water surface: water flow is primarily above ground (Alexandria Water Pollution Control Federation, 1990).

Source: Sinclair, 2000

Figure 1: Surface Flow Wetland Type

1.3.2 Subsurface Flow Wetland

The Subsurface flow (SSF) wetland, which is originated in Europe over 20 years ago (Joseph, 2005), consists of a sealed basin or channel with a porous substrate of rock or gravel media and a barrier to prevent seepage. The media also support the root structure of the emergent plants. The design of these a system assumes that the water level in the bed will remain below the top of the rock or gravel media (USEPA, 1993; Luise *et al.*, 1999; Martha, 2003) (Figure 2).

The treatment processes in SSF wetland system is more efficient than in the SF wetland system; because the media provides a greater number of small surfaces, pores and crevices where treatment can occur. Waste consuming bacterial attach themselves to the various surfaces, and waste materials in the water become trapped in the pores and crevices on the

7

media and in the spaces between media. Chemical treatments also takes place as certain waste particles contact and react with the media (USEPA, 1993)

The biological treatment in SSF wetlands is mostly anaerobic, because the layers of media and soil remain saturated and unexposed to the atmosphere (Sinclair, 2000). However, wetland plants are able to grow extensive roots even in these anaerobic conditions. The area where the roots grow is called root zone and usually includes the upper 0.15 to 0.40 meter of media. If cells are alternated or allowed to rest periodically, or if the water level is regularly cycled, the roots can reach throughout the media layer (Pottir and Karathanosis, 2001)

According to Brix, (1994), wetland plant roots contribute oxygen to the cells which allow some aerobic treatment to take place in the root zone, which stimulates organic matter oxidation and the growth of nitrifying bacteria. This is because vascular wetland plants are equipped with *aerenchyma* or *aerenchymous* tissues containing a network of tiny hollow tubes that traverse the length of the plant allowing gases to move from the above water part of the plant to the roots and rhizomes, and vice versa (Richard *et al.*, 1994) .

In addition, these plants have *lenticels,* or small openings along the plant stems that facilitate the flow of gases in and out. Lenticels may also be located on adventitious roots that develop from the stalk or stem of the plant within the water column (Kandlec and Knight, 1999). Other structural components include "knees" on Cyprus trees (an emergent woody plant) and buttresses, also on certain woody species (Luise *et al.*, 1999). Generally, the transfer of gases and in particular of oxygen from the above-water part of the emergent herbaceous plants to the root zone can occur in two basic ways; passive molecular (gas-phase) diffusion and bulk flow of air through internal gas spaces of the plant, (resulting from internal pressurization).

Plants further contribute to wastewater treatment by providing additional surfaces where bacteria can reside and where waste materials become trapped (Faithful, 1996; Joseph, 2005). Plants also take-up and store some of the metals and nutrients in the wastewater. Most subsurface flow wetlands are designed so that wastewater travels through the length of the cell one time to receive treatment. Typical retention times range from two to five

8

days for BOD_5 removal and seven to fourteen days for nitrogen and phosphorus removal (Crites and Tchobanoglous, 1998). Due to this, SSF wetlands have most frequently been used to reduce BOD_5 from domestic wastewaters (Godfrey *et al.*, 1985; Rechard, 1998)

SSF CW cells are usually designed with aspect ratios, (length to width ratio), of 3:1 or less (USEPA, 1998; Simi and Matchell, 1999). Wider cells tend to be more cost effective because long narrow cells must be deeper and require more treatment media. In addition, the wastewater is less likely to back up in wider cells if too much water enters the system (Luise *et al.*, 199).

The SSF type of CW, which was the most important concern of this study, is thought to have several advantages over the SF type if the water surface is maintained below the media surface there is little risk of odors, and insect vectors (Vymazal, 2002). In addition, it is believed that the media provides greater available surface area for treatment than the SF concept so the treatment responses may be faster for the SSF type, which therefore can be smaller in area than a SF system designed for the same wastewater conditions (Wallance, 1998; USEPA, 1993).

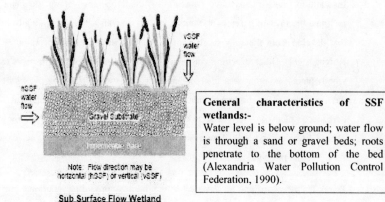

General characteristics of SSF wetlands:-
Water level is below ground; water flow is through a sand or gravel beds; roots penetrate to the bottom of the bed (Alexandria Water Pollution Control Federation, 1990).

Note: Flow direction may be horizontal (hSSF) or vertical (vSSF)

Sub Surface Flow Wetland

Figure 2: Subsurface Flow Wetland Type

Source: Sinclair, 2000

2. LITERATURE REVIEW

2.1 Performance Evaluation of Constructed Wetlands

The use of CW for wastewater treatment was stimulated by a number of studies in the early 1970s that demonstrated the ability of wetlands to remove suspended sediments and nutrients in wastewaters (Nichols, 1983; Godfrey *et al.*, 1985; and Knight, 1990). There are three major important nutrients that are commonly found in municipal wastewaters; nitrogen (N), phosphorus (P), and potassium (K). But the concentration of K is not taken into consideration since K holds no health risk, is not typically present in wastewaters in the optimum combination with N and P, and is often found in great abundance naturally in nature (William and James, 1993; Reed *et al.*, 1995).

Untreated domestic wastewater usually contains among other contaminants, nutrients mainly nitrogen and phosphorus that can stimulate the growth of aquatic plants which in turn results in various environmental pollution related problems. For this reason the treatment of wastewater is not only desirable but also necessary (Knight, 1990). Treatment is necessary to correct wastewater characteristics in such away that the use or final disposal of the treated effluent can take place in accordance with the rules set by the relevant legislative bodies without causing an adverse impact on the receiving water bodies (Njanu and Mlay, 2003).

Thus the two objectives of wastewater treatment, separating wastes from wastewater and preventing pollution of the receiving waters are evaluated differently. Treatment efficiency depends on the extent to which specific waste materials are separated from the wastewater and can be calculated for a number of different Parameters (Christon, 2004).

For domestic wastewater treatment, the pollutants of most concern are biochemical oxygen demand (BOD^5), chemical oxygen demand (COD), total suspended solids (TSS), total nitrogen (TN), ammonium nitrogen (NH^+_4-N), nitrate nitrogen (NO_3^-N), orthophosphate(PO_4^{3-}), total phosphorus (TP), sulfate (SO_4^{2-}), sulfide (S^{2-}) and Sanitary indicators-(total Coliforms or fecal Coliforms) (Wallance, 1998; USEPA, 1993;Aimee, *et al.*, 2000). Specification of the wastewater components measured or the form of pollution evaluated is essential for describing the efficiency and effectiveness of wastewater treatment plant. Most often, removal efficiency is expressed as a percentage and these values are used to compare different treatment processes and to determine if a particular treatment plant is accomplishing that for which it was developed (Christon, 2004)

The major function of wastewater treatment plant is to reduce the organic loading of domestic wastewater so that it can be safely discharged to the receiving stream. The effectiveness of the sedimentation process is monitored through BOD_5, COD and TSS parameter (Smith, 1999). Results of a research conducted in Northern Alabama by Kathleen (2000) showed that constructed wetlands can reduce the organic content, appreciably to an average of 85 percent. Kathleen also showed by her study, the effluent BOD_5 was below the required (30mg/L), which was under 22mg/L.

In addition to this, TSS removal is very effective in SSF CW systems. Most of the removal probably occurs within the first few meters of travel distance from the inlet zone (Smith, 1999). It would appear that with HRT of a day, the TSS will be removed to a level of about 10 mg/L (USEPA, 1993). Six constructed wetland projects which were found in Listowel, Santee, Sidney, Arcata, Emmetsburg and Gustine, were evaluated for their performance on the removal of TSS, the result showed that the treatment plants removal efficiencies were; 93, 90, 92, 28, 73 and 86 percents, respectively (USEPA, 1988).

A study conducted in Kenya to assess the effectiveness of CW in treating domestic wastewater showed that the removal of BOD_5, TSS, COD, TN, NH_4-N and Orthophosphate were highly effective with a removal value of 98%, 85%, 96%, 90%, 92%, and 88%, respectively (NyaKango and VanBruggen, 1999). This was mainly because this wetland consists of a combination of a SF system followed by three SSF wetland cells in a series adjacent to it.

There is substantial evidence in the design of CW that a number of cells in series can consistently produce a higher quality effluent. Because this process minimizes the short circuiting effects of any one unit and maximizes the contact area in the subsequent cell (Gearheart, 2004). The series cells allow for a wider range of pollutant removal as well as allowing for the effective removal of fractions of some contaminants i.e. dissolved inorganic, particulates and organic forms (Faithful, 1996).

11

Because of this, it is generally recommended that for treatment and water quality purposes CW should consist of a minimum of 2 to 3 cells in series (Gearheart, 2004). In addition, this study indicated that if all cells of a CW are planted with different plant species it produces good quality effluent; since multiple plant species within the system maximized root biomass in the wetland substrate resulting in more efficient treatment (Oketch, 2003).

A case study conducted in Italy, to assess the treatment performance of a SSF CW by Pucci *et al.* (2000), has showed high removal efficiencies for COD (93%), TSS (81%), hygienic parameters(TC 99%, FC 99.7%), but relatively low for nitrate (55 %), total nitrogen (50%)and ammonium (30%), very low for total phosphorus (20%). This is mainly due to poor nitrification and denitrification in the system.

Microorganisms play a crucial role in the transformation and removal of nitrogen in wastewaters. Organic nitrogen is transformed in to NH_4^+-N through a complex biochemical process called ammonification. This process is performed mainly within the root system. The NH_4^+-N produced from this process is preferred for the formation of biomass in growing plant (Kandlec and Knight, 1999)

NH_4^+-N produced by ammonification can then be converted in to NO_2^--N (Nitrite nitrogen) and NO_3^--N by microbial nitrification. NH_4^+-N is transformed first in to NO_2^--N, which is unstable and then in to the chemically stable NO_3^--N. Nitrate nitrogen can be used as a nutrient for plants and may play a role in eutrophication. Nitrification occurs in aerobic conditions with a pH of at least 7.2 units. Denitrification is achieved in anoxic conditions in which nitrate or nitrite serves a respiratory electron acceptor for denitrifying bacteria to carryout the oxidation of carbonaceous organic substrates, with a pH range of 6.5 to 7.5 units (Gersberg *et al.*, 1984; Kadlec and Knight, 1996; Simi & Mitchell, 1999; Brix *et al.*, 2003). The fate of nitrogenous wastes in constructed wetland is summarized in Figure 3 below.

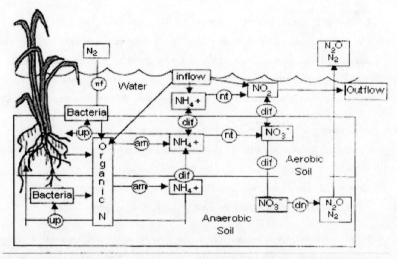

Figure 3: Generalized diagram which shows transformation of nitrogen in constructed wetland systems.
Where am = ammonification, nf = nitrogen fixation, dif = diffusion, nt = nitrification, dn = denitrification, up = uptake

Source: Faulkner, 2004

In addition to nitrogenous waste, domestic wastewaters contain phosphate from cleaning products; because of this it contains high concentration of phosphorus than other wastewater sources (Renee, 2001; Martha, 2005). When determining the role of phosphorus retention by wetlands, particularly wetland substrates, it is important to understand the forms of phosphorus in the system and that have ecological and environmental consequences (Faulkner, 2004).

In wastewater entering a wetland particularly secondarily treated effluent from a sewage treatment plant, (Septic tank), the Phosphorus component will be composed predominantly reactive phosphate (mainly orthophosphate) (Miriam *et al.*, 2002) and total phosphorus content comprises both inorganic and organic particulate and filterable non-reactive phosphorus forms (Kathleen, 2000). The forms of phosphorus and its transformation in wetland systems is summarized in Figure 4 below

13

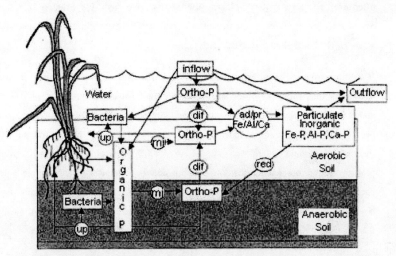

Figure 4: Generalized diagram which shows phosphorus transformations in constructed wetland systems.
Where mi = mineralization, ad/pr = adsorption and precipitation reactions with Fe, Al and Ca, dif = diffusion, red = Fe-reduction, up = uptake

Source: Faulkner, 2004

Because of the limited contact opportunities between the wastewater and the soil, phosphorus removal in most CW systems is not very effective (USEPA, 1988; Reddy, 2004). For example, a gravel media with its high conductivity permit all of the water to flow within the bed but because of the impermeable nature of the bed have only a limited surface area for adsorption, ion exchange and/or chemical reaction to take place because of this, once the active sites are utilized, phosphorus removal ceased (Newman *et al.*, 2000). Some systems in Europe use sand (clay) instead of gravel to increase the phosphorus retention capacity, but selecting this media results in a larger system because of the reduced hydraulic conductivity of sand compared to gravel (USEPA, 1993.)

The results of the investigation done by Miriam *et al.* (2002), on the efficiency of phosphorus retention in CW treating agricultural drainage water for two years (2001-2002), suggested that a minimum HRT to retain at least 50% of the bio-reactive phosphorus was 7 days. These researchers also confirmed that if the HRT exceeded 10 days, the removal efficiency ranged from 50 to 90%, but decreased drastically and was

14

often even negative, if the HRT was shorter than five days. For example, the wetland assessed by this study indicated that, it only retained 2% of the bio-available phosphorus, since the HRT was shorter than seven days.

In addition to nitrogen and phosphorus, sulfur occurs in CW in different forms, but because of their impacts, sulfide and sulfate are the most important in wetlands (William and James, 1993; Renee, 2001). In the wetland systems, sulfate reduction occurs due to the presence of sulfate reducing bacteria in the substrate coupled with sufficient organic material to stimulate their activity (Martha, 2003). These bacteria are a group of prokaryotic microorganisms that use electron donors to reduce sulfate. Sulfate reducing bacteria remove sulfate from the water column by metabolizing sulfate into living tissue or by reducing sulfur to produce energy (Hsu, 1998; Simi and Mitchell, 1999). Forms and the fate of sulfur in CW are summarized in Figure 5.

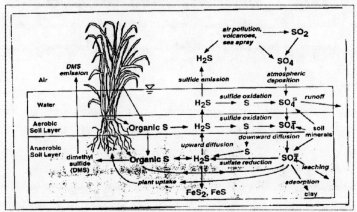

Figure 5: Generalized diagram which shows Sulfur transformations in constructed wetland systems.

Source: William and James, 1993

Evidences for the presence of sulfate reducers in wetlands include blackened sediment due to the resulting precipitation of iron sulfides and the release of sulfides cause the odor familiar to those who carry out research in wetlands, 'the smell of rotten eggs'(William and James, 1993; Fauque,1995). Although it is rarely present in such low concentrations that is limiting to plant or animal growth in wetlands, the hydrogen sulfide that is

15

characteristic of anaerobic wetland sediments can be very toxic to plants and microorganisms, especially when the concentration of sulfate is high (Morse *et al.,* 1987)

A study conducted by Fauque, (1995) in New York to assess the efficiency of five constructed wetland for sulfate removal, indicated that sulfate reduction occurred in all five wetlands which varies in degrees of treatment effectiveness from 54–70%. According to his conclusion, this difference in removal efficiency was mainly due to the difference in plant species and the substrates used in each wetland.

The other most important wastewater parameter, especially with regard to human health, is the removal of coliform bacteria (Gersberg *et al.,* 1984), which in turn indicates the removal of pathogenic microorganism from the wastewater (Kathleen, 2000; Pucci *et al.,* 2000; Christon, 2004). The results of these researchers' work showed that on average coliforms bacteria removal in CW was greater than 90 percent.

In addition to these, the result of a study conducted by Mariade'j *et al.,* (2001) demonstrated that a significant removal (more than 90%) of indicator microorganisms can occur in CW receiving domestic wastewater with only 1 to 2 day detention time. But this high percentage removal of indicator bacteria does not always equate with producing outflow with acceptable level. For instance Tanner and Sukias (2003) reported their findings of fecal coliform removal efficiency of 30% to 85% and suggested that a reduction below 300-500 cfu/100ml was difficult to obtain.

Generally, while constructed wetlands have such a proven effectiveness for treatment of a variety of wastewaters in developed countries, little work has been done in developing countries like Ethiopia where the concept of constructed wetlands for wastewater treatment is still a relatively new idea (Kaseva, 2003; Muhammad *et al.,* 2004). Even it is not known by many stakeholders who can be beneficiary in using this method to treat their waste water.

2.2 The Contribution of Wetland Plants for Wastewater Treatment

The presence of wetland plants has been hypothesized to play a key role in wastewater remediation (Luckeydoo and Fausey, 2002). In addition to their aesthetic roles, wetland plants exhibit several properties which enhance wastewater treatment processes and thus make them an essential component of the treatment wetland. These properties influence wastewater treatment through physical effects such filtration, adsorptions followed by sedimentation, and provision of surface area for the growth and attachment of microorganisms (Sinclair, 2000).

Metabolically, plants take up pollutants; produce organic carbon and oxygen, there by improving the water to varying extents (Joseph, 2005). Plants in wetland systems have been viewed as storage compartments for nutrients where nutrient uptake is related to plant growth and production. Emergent plants utilize their roots to obtain sufficient nutrients from wastewater. Free floating species have roots with numerous root hairs and successfully obtain nutrients from both the water column and substrate (USEPA, 1988).

They often grow in gravel beds to stimulate uptake and create suitable conditions for the oxidation of the substrate, there by improving the ability of the system to treat wastewater (Njanu and Mlay, 2000). This needs consideration of plant selection and management techniques that create rhizosphere surface area per volume of bed and bed design; optimal depth, HRT and media of constructed wetland (Muhammad *et al*, 2004).

If the wetland plant is intended as a major oxygen source for nitrification in the system, then the depth of the bed should not exceed the potential root penetration depth for the plant species to be used. This will ensure availability of some oxygen throughout the bed profile, but may require management practices which assure root penetration to these depths (USEPA, 1993, Gersberg et al., 1998)

From the standpoint of wastewater treatment, certain plant species appear to be more efficient in CW treatment systems and others may be more tolerant of high pollutant concentrations. It appears that major contribution from the vegetation in SSF system is service of the root/rhizome structure as a substrate for microbial activity and is a limited oxygen source for nitrification (Njanu and Mlay 2000)

17

Plant species selection can have impacts on sedimentation, plant nutrient accumulation, and the creation of microenvironment that facilitates microbial degradation of contaminants (Luckeydoo and Fausey, 2002). Further more, plant species selected for constructed wetland cells shall be hydrophyt plants suitable for local climatic conditions and tolerant of the concentration of nutrients and other constituents in the wastewater stream and selected for their treatment potentials. Preference shall be given to native wetland plant materials collected or grown from materials collected within a 323 kilometers radius and plant species collected within this radius are considered local origin (USEPA, 1988; Indian Natural Resources Conservation Service, 2001).

Taking this fact into consideration, JWBO's CW is covered with three different plant species, identified in the local areas, namely *Cyprus papyrus*, *Cyprus alternifolia* and *Phoenix canariensis*, which are among the typical characteristic species of wetland ecosystems of Ethiopia (EIBC, 2007). During the assessment time conducted in this study period, it was confirmed that these plant species are common in different wetlands of Ethiopia such as Lake Tana, Lake Awassa, Lake Zeway, and in different Lakes found in Debere-Zeyet.

The *Phoenix canariensis* commonly called palm, is very widely planted as an ornamental plant in warm temperate regions of the world. It is cultivated as a street tree in many of the larger towns and cities with altitude of 1000–2400 m.a.s.l. (Sebsebe *et al.*, 1997) particularly in areas with continental climates where temperatures never fall below 10 °C. (http://en.wikipedia.org/wiki/Canary_Island_Date_Palm accessed on 12 May 2007).

It is also common in most parts of Ethiopia as ornamental plant in cities such as Addis Ababa, Bahir-Dar, and Jimma and in natural wetlands found in different parts of the country (Sebsebe *et al.*, 1997; EIBC, 2007). In addition to these, this plant is used by the local community to make a variety of decorative handicrafts, such as baskets in different parts of Ethiopia (Afework, 2006). These palm attracts many birds particularly pink breasted pigeon to nest in it. The fruits are sweet and much liked by children (Sebsebe *et al.*, 1997)

18

Cyprus papyrus, commonly called papyrus, is a member of the sedge family (Cyperaceae). It is a monocot that is native to river banks and other wet soil areas in Egypt, Ethiopia, the Jordan River Valley and other parts of the Mediterranean basin. Throughout the world these plants hold great regional importance in weaving mats, baskets, screens and even sandals (Matt, 1997; Sebsebe *et al.*, 1997).

According to Afework, (2006), even though it is over utilized and it is at risk currently, in Ethiopia for 'Woyto' community (a particular ethnic group found around Lake Tana) and other local people, Papyrus is an important raw material used for craft making and ceremonial purposes. In addition to this, since the roots of Papyrus is spread over the water forming floating mat, helps to prevent soil erosion and trap polluted sediments in the wastewater. A study conducted by Abe, Ozaki and Kihou (1997) and Joseph, (2005) showed that *C. papyrus* is useful in wastewater treatment. This study showed that *Cyprus papyrus* reduced the amount of nitrogen and phosphorus in wastewater by more than fifty percent.

Cyprus alternifolia commonly called Umbrella plant, is a semi-aquatic and requires a very moist soil and a medium light exposure for growth. It is also cultivated as a garden plant, within 700–2400 m.a.s.l (Sebsebe *et al*, 1997). During the assessment done in this study period, *C. alternifolia* is common in natural wetland found in Central Refit Valley of Ethiopia, such as in Lake Awassa, Lake Zeway and in different Lakes found in Deber-Zeyet.

Additionally, this plant species was more common in natural wetlands found in other parts of Ethiopia. Like that of *C. papyrus*, it is also over utilized and currently it is at risk. For example in Western Gojjam, (Dangella, Achefer, and Mecha districts), this plant species was used by the local communities to produce materials that protect themselves from rain. Since most wetlands in these districts are changed in to farmlands and others are overgrazed, they have stopped to make such type of material; instead they are using thin plastics to protect themselves from rain.

19

Though the socio-economic advantages of these wetland plant species are well known, their potential for wastewater treatment is not well understood. For instance, only few studies investigated the potential of *C. papyrus*, (which colonizes many wetlands in Africa), for wastewater treatment (Joseph, 2005), but no published information has been found during the preparation time of this study on the potential of *C. alternifolia* and *P. canariensis* for wastewater treatment.

Therefore, there is a need for further research on the wetland plant species adapted to the local ecological conditions of Ethiopia in order to supplement and optimize the treatment efficiency of constructed wetlands.

3. RATIONALE

Environmentalists have referred to wetlands as nature's kidneys. Much interest has developed in recent years in using CW to remove contaminants from water, whether it is effluent from domestic, municipal, industrial, agricultural wastewaters, or acid-mine drainage (Kenneth, 2000). While constructed wetlands have proven their effectiveness for treatment of a variety of wastewaters, this technology is new concept in Ethiopia. Only very few institutions such as Jehovah's Witnesses Branch office is using it to treat their domestic wastewater. Even this institution did not know the treatment efficiency of this system until this study was conceived.

This is mainly because of the lack of information, research work and appropriate design for the effectiveness of CW to treat wastewater. The primary purposes of this study was, therefore, to evaluate the effectiveness of constructed wetland by taking JWBO CW as the case study and to recommend that constructed wetlands can be an alternative wastewater treatment technology.

The efficiency of CW varies with site specific parameters such as pH, wastewater temperature, Hydraulic Retention Time and Hydraulic Loading Rate (Kenneth, 2000; Renee, 2001). In addition to these, the original wastewater contaminant concentrations impacts wetland treatment efficiency (Faithful, 1986; Joseph, 2005) wetland construction should therefore be altered based on the original wastewater characteristics (Woulds and Ngwenya, 2006). For example, a study conducted in Scotland (2006), highlighted the importance of site-specific conditions in regulating the effectiveness of a constructed wetland for domestic wastewater treatment. So, this study was conducted to confirm these site specific conditions.

In addition to treatment, alternatives of effluent management options must be considered as part of any CW system (John and Partick, 2000). The disposal of effluent needs to include proper design elements and proper regulation approval. Some of the options available include; infiltration strips, grass filter strips, recycling back through the wetland, irrigation and direct discharge (Kenneth, 2000).

Of these effluent management options, the one which is practiced by JWBO was direct discharge to the surface water (nearby stream). While simple, it requires the most attention to regulatory testing and monitoring to ensure that the national pollutant discharge elimination system requires or more stringent local standards are met. The purpose of this study was also to check whether the effluent quality meets the provisional discharge limit values of Ethiopia set by National Environmental Quality Standards for domestic wastewater effluent (EEPA, 2003) to discharge in to surface and ground water bodies.

Generally, the following points are indicated as the primary significance of this study:-

➢ Helps to determine the efficiency of constructed wetland for domestic wastewater treatment, determines the effluent quality and recommends the possible effluent management options of JWBO

➢ Based on this treatment efficiency, the site serves for educational/research purposes as alternative wastewater treatment technology that works in more a natural manner.

➢ The result of this study also helps to promote for more widespread use of this technology which was both energy and cost effective alternative wastewater treatment system in the country with the added benefits of providing wildlife habitat and recreational value in the country.

➢ In addition, this study helps to document the contributions of different plant species for wastewater treatment process of constructed wetlands.

4. OBJECTIVES

4.1 General Objective

The general objective of this study was to evaluate the treatment performance of constructed wetlands as an alternative municipal wastewater treatment technology under Ethiopian climatic conditions.

4.2 Specific Objectives

- To determine removal efficiency of constructed wetlands for selected wastewater quality parameters; BOD_5, COD, TSS, NH_4^+-N, NO_3^--N, TN, PO_4^{3-}-P, TP, SO_4^{2-}, S^{2-}, and Coliform bacteria of domestic wastewater, taking JWBO CW as a case study,

- To evaluate the removal capacity of different wetland cells planted with *C. papyrus, C. alternifolia* and *P. canariensis* at JWBO CW system,

- To recommend appropriate design and plant species as well as effluent management options based on the quality of the effluent from the wetland, and

- To provide the necessary information for local, regional and national governments for wider application of the technology.

5. MATERIALS AND METHODS

5.1 Description of the Study Area

This study was conducted on CW found at Jehovah's Witnesses Branch Office (JWBO) located in South-West part of Addis Ababa, Ethiopia, which was operating since 2004. The wastewater generated in different parts of the building is delivered in to the septic tank, with $148.72m^3$ (length 14.30m * width 4m * depth 2.6m) capacity. The primary treatment of the wastewater is done in this septic tank. It contains baffles and other screening materials to prevent the solid materials leaving the settling tank. Then the wastewater enters to the equalization tank, with 36.98 m^3 (Length 4.4m * Width 4.2m * Depth 2.0m) capacity.

The equalization tank provides a multipurpose benefit when considering water budget: the amount of water going in to, flowing out and remaining in the wetland. It will serve as additional settling basins for removing solids as well as storage for excess water during high water input (Kenneth, 2000). In addition to these, this tank contains a gravity flow pumps that releases the influent to the wetland when the depth of the water reaches 1.46m with uniform hydraulic loading rate ($17.74m^3$) and which happens on average two times per day.

The purpose of such pump is to discharge the contents intermittently in to the wetland. This intermittent discharge permits the filtering material to be completely dosed with a rest interval following after each application of wastewater, thus prolonging the usefulness and efficiency of the wetland system (USEPA, 1993). The discharge from the equalization tank assures completely filling the cells, thus insuring that every part of the cell filled will be utilized effectively.

This SSF wetland is consisted of six cells with the area of $98m^2$ (Length 14m and width 7m) each and working in parallel (Figure 6). The total area of JWBO constructed wetland is $588m^2$ (length 42m and width 14m). The depth of the wetland is 0.70m, filled with gravel having different sizes (40-80mm and 20-30mm) as a substrate. Each cell received equal volume of influent from the equalization tank and it was regulated by the gate valves of each cell (Figure 8). The effluent of each cell is collected with the help of perforated

24

pipe and released in to the collection ditch. All these collected effluents were transported and disposed to the nearby stream.

The wetland, in which this study was conducted, was populated with three plant species, namely *Cyprus papyrus (*Papyrus), *Cyprus alternifolia* (Umbrella plant) and *Phoenix canariensis* (Palm*)* (Figure 7). Of these, *C. papyrus* and *C. alternifolia* were obtained locally at Debre-Zeyet, which is approximately 47 Kilometers from the wetland site, and the other was obtained in Addis Ababa. One plant species was planted in one wetland cell and the species are planted in an alternative way throughout the wetland system. As indicated in Figure 6 and 7; Cell 1 and 3 were planted with *C. papyrus*, cell 2 and 5 were planted with *C. alternifolia* and cell 4 and 6 were planted with *P. canariensis*

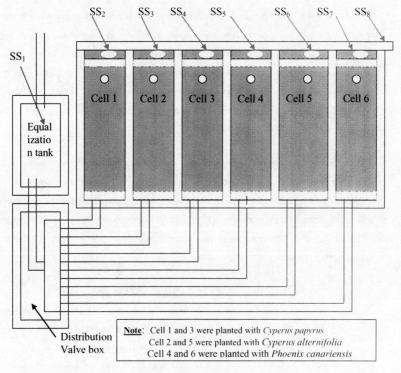

Figure 6: Sketch map of JWBO Wetland and sampling sites for this study

25

C. alternifolia; Cell 2 and 5	C. papyrus; Cell 1 and 3	P. canariensis; Cell 4 and 6

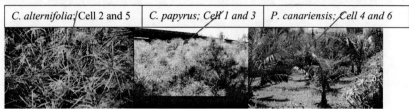

Figure 7: Wetland Cells and the type of Plant species planted at JWBO CW.

A. partial view of JWBO   **B. partial view of distribution valves**

Figure 8: Partial view of JWBO and wastewater distribution valves of the wetland

5.2   Sampling

Before sample collection, the hydraulic retention time (HRT) of JWBO CW was calculated. This is because wastewater treatment processes are dependant. amongst other factors; on the period of time that wastewater physically resides within the wetland. The approximate estimate of HRT for this wetland was obtained by using Darcy's formula (USEPA, 1993)

$$HRT = \frac{nLWd,}{Q_{av.}} \quad \text{----------------- Darcy's Law}$$

Where:

n = effective porosity media, % as a decimal.

L = Length of the bed, (m)

W = Width of the bed, (m)

d = Average depth of liquid in bed, (m)

Q_{av} = the average of the inflow and outflow $\frac{\{Q_i + Q_o\}}{2}$, (m^3/day),

Source: USEPA, 1993

26

The porosity (n) is used to determine the actual flow velocity in the void spaces in retention time calculation equation. Porosity is equal to void volume/total volume and is expressed as percentage and it is 35% for gravel with the size of 80mm, (USEPA, 1993)

The quantity of inflow to the wetland was determined by measuring the depth of the water in the equalization before the pump was opened and immediately after the pump was closed, (which was uniform in hydraulic loading rate) and the outflow quantity of the wetland was calculated manually using graduated container with a stopwatch.

Based on this, the mean inflow and outflow of the wetland was 35,480 litters (35.48m3) and 27,300litters (27.3m3) per day, respectively. The mean depth of water level of the wetland, (0.33 meter), was calculated by measuring the water levels through the water level monitoring pipe found in each cell of the wetland using a measuring tape. Based on the above data, the calculated hydraulic residence time (HRT) of the wetland was 2.16 days.

To evaluate the performance efficiency of CW, inlet grab samples were collected two days before the outlet ones, according to the estimated hydraulic retention time of the wetland. As indicated from Figure 6 and 9, samples were collected at eight sites (SS_1, SS_2, SS_3, SS_4, SS_5, SS_6, SS_7 and SS_8). SS_1 was in the equalization tank (for influent of the wetland), while SS_2 to SS_7 were in manholes of each wetland cells (for effluent of each wetland cells) and SS_8 was at the final disposal site (for effluent of the overall wetland system). In all these sites, samples were collected for two months (May 14 to July 14, 2007). A total of 24 samples (three samples per site) were collected in every 15 days throughout the study period.

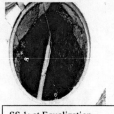

SS 1: at Equalization tank	SS 2 – SS 7: at effluent of each cell	SS 8: at the overall wetland system effluent

Figure 9: Sample sites at JWBO CW systems in which all samples were collected.

For analyses of physicochemical parameters, samples were collected using 500 ml plastic bottles washed with distilled water and repeatedly rinsed with the wastewater at each sample site before the sample was collected. For coliform bacteria tests, sample collection was carried out using glass bottles previously sterilized by autoclave. Additionally, for plant species identification, three vegetation samples were collected from cells planted with different plant species (Figure 10)

| Planted at cell 2 and 5 | Planted at cell 1 and 3 | Planted at cell 4 and 6 |

Figure 10: Wetland plant sampling sites at JWBO CW

5.3   Analysis

A detailed influent and effluent characterization of all collected samples was carried out for both selected physicochemical and bacteriological wastewater quality parameters. The common influent and effluent quality parameters that were determined were: BOD_5, COD, TSS, ammonium N, nitrate N, total N, orthophosphate, total phosphorus, sulfate, sulfide, TC, FC, wastewater pH and temperature.

COD, ammonium N, nitrate N, total N, orthophosphate, total phosphorus, sulfate and sulfide were measured calorimetrically by spectrophotometer (DR/ 2010, USA) according to HACH instructions. BOD_5 was determined using standard methods of APHA (1998); Total Suspended Solids was determined using gravimetric method while Wastewater temperature and pH was measured on-site using portable thermometer and pH meter.

The bacteriological quality indicators; Total Coliform (TC) and Fecal Coliform (FC) were evaluated using Membrane Filter (MF) procedures of standard method for the examination of water and wastewater (APHA, 1998). Samples were serially diluted (10^1 to 10^6) using double distilled water and 100 ml of the diluted water was filtered through a filter paper in order to retain bacteria using filtering unit. Then the filter paper was placed on a surface of absorbent pad socked with Membrane Lauryl Sulfate Broth and incubated at 37°c for Total Coliforms and 44°c for Fecal Coliforms, Yellow Colonies, which is typical colony characteristic of TC and FC using Membrane Lauryl Sulfate Medium (APHA, 1998), (Annex III: Figure 1 & 2), were counted using colony counter and the results were recorded as the number of Colony Forming Unit (CFU) of TC and FC per 100 ml. Additionally, plant species classification identifications was done in the National Herbarium of Ethiopia

The product of the influent hydraulic discharge data and the nutrient concentration obtained in the influent divided by the total area of the wetland gives the nutrient loading rate in the wetland (Healy and Cawley, 2001). The removal efficiency of the wetland for each wastewater quality parameters were calculated using the following formula:

$$\text{Efficiency (\%)} = \frac{[C_i - C_e]}{C_i} * 100$$

Where: C_i = is the concentration of the waste material in the influent

C_e = is the concentration of the waste material in the effluent

Source: Christon, 2004

To determine the contribution of wetland plant species; the results obtained from the effluent samples of constructed wetland cells covered with *C. papyrus* (Cell 1 and 3), *C. alternifolia* (Cell 2 and 5) and *P. canariensis* (Cell 4 and 6) plant species were compared against each other.

5.4   Statistical Analysis

Statistical analysis was performed with SPSS and MINITAB packages Release 13.00 for windows. The included Mean, Standard Error and Pearson correlation analysis was done using SPSS package and MINTAB package was used for Analysis of Variance (ANOVA) test.

6. RESULTS AND DISCUSSION

To evaluate the treatment performance of JWBO CW, selected parameters: BOD_5, COD, TSS, NH_4^+ N, NO_3^- N, TN, PO_4^{3-}, TP, sulfate, sulfide, pH, wastewater temperature and coliform bacteria (TC and FC) were measured in wetland influent and effluent. During the entire period of the study a total of 24 samples were analyzed for each wastewater quality parameter. The mean influent and effluent concentrations as well as the changes in concentration i.e. percentage removals of the wetland for the selected parameters were presented as follows:

6.1 BOD_5, COD and TSS Removal

Table 1 presents the mean influent and effluent concentrations of BOD_5, COD and TSS for each wetland cells and the overall system of JWBO CW. The average daily inflow rate of the wetland during the study period was 35,480 liter per day. The average influent wastewater temperature and pH values were 25.8 ± 0.3 ^0C and 7.1 ± 0.27 pH units respectively. The mean influent parameter values were: BOD_5 (273.3 ± 21.9 mg/L), COD (619.3 ± 187.3 mg/L) and TSS (201.3 ± 6.7 mg/L), which was equivalent to the mean daily loading rate of 16.5 $kg/m^2/day$, 37.4 $kg/m^2/day$ and 12.2 kg $/m^2/day$, respectively. Since each wetland cells receive equal volume of wastewater from the equalization tank, the mean influent values were the same.

The mean effluent parameter values of JWBO CW were: BOD_5 (2.0 ± 0.6mg/L), COD (68 ± 11.7 mg/L) and TSS (30.0 ± 10.5 mg/L). The mean pH value was 7.18 ± 0.013 with a range of 7.15 – 7.19, which is nearly neutral. Figure 11 shows the average removal efficiency of each wetland cells and the overall wetland system of JWBO CW. The average removal efficiency of each wetland cells were within the range of 97.3% - 98.4% for BOD_5, 87% - 89.8% for COD and 81.5% - 84% for TSS (Table 1 and Figure 11). The overall JWBO CW removal efficiency was: BOD_5 (99.3%), COD (89%) and TSS (85%).

Table 1: **Mean BOD$_5$, COD and TSS Influent & Effluent concentration values (mg/L) of JWBO CW**

Wetland cells	BOD$_5$			COD			TSS		
	Influent	Effluent	%	Influent	Effluent	%	Influent	Effluent	%
Cell 1[a]	273.3 ± 21.9	6.3 ± 1.9	97.7	619.3± 187.3	63.7± 14.5	89.7	201.3 ± 6.7	32.7 ± 11.4	83.8
Cell 2[b]	273.3 ± 21.9	7.3 ± 2.3	97.3	619.3± 187.3	76.3± 17.6	87.7	201.3 ± 6.7	37.3 ± 10.2	81.5
Cell 3[a]	273.3 ± 21.9	5.3 ± 0.9	98.0	619.3± 187.3	66.7± 15.2	89.2	201.3 ± 6.7	32.3 ± 12.0	84.0
Cell 4[c]	273.3 ± 21.9	6.0 ± 2.6	97.8	619.3± 187.3	63.3± 11.2	89.8	201.3 ± 6.7	33.0 ± 7.2	83.6
Cell 5[b]	273.3 ± 21.9	5.0 ± 0.6	98.2	619.3± 187.3	80.3± 20.5	87.0	201.3 ± 6.7	34.0 ± 7.2	83.1
Cell 6[c]	273.3 ± 21.9	4.3 ± 0.9	98.4	619.3± 187.3	65.0± 11.5	89.5	201.3 ± 6.7	34.7 ± 7.5	82.8
Overall wetland system	273.3 ± 21.9	2.0 ± 0.6	99.3	619.3± 187.3	68.0± 11.7	89.0	201.3 ± 6.7	30.3 ± 10.5	85.0

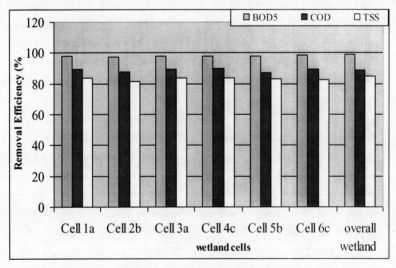

Figure 11: BOD$_5$, COD and TSS Removal Efficiencies of JWBO CW

[a] = wetland cells planted with *C. papyrus* species
[b] = wetland cells planted with *C. alternifolia* species
[c] = wetland cells planted with *P. canariensis* species

31

This finding was similar to that of the study done in USA: USEPA (1988) BOD_5 (93%); in Kenya: Nyakango and VanBruggen (1999) BOD_5 (98%), COD (96%) and TSS (85%); in Northern Alabama: Kathleen (2000) BOD_5 (85%); in Italy: Puccie *et al.* (2000) COD (93%) and TSS (81%); and in Tanzania: Kaseva (2003) COD (33.6 – 60.7%) removal efficiency using subsurface constructed wetland

Biochemical oxygen demand (BOD_5) is widely used as a measure of the polluting potential of effluent (discharge). It is a measure of the oxygen used by bacteria in breaking down the organic biodegradable load (fats, proteins, carbohydrates, etc) in sewage discharges. The COD value provides an indirect indication of the potential oxygen depletion that may occur from the discharge of organic materials in surface waters. As a result BOD_5 and COD removal efficiencies are the major criteria for the selection of wastewater treatment facilities.

The result indicated that BOD_5 and COD removal efficiency of JWBO was very efficient with a removal value of 99.3% and 89%, respectively. This showed the effectiveness of constructed wetland systems to remove organic matter in domestic wastewater. The high removal efficiency might be due to the fact that organic matters in domestic wastewater are dominated with readily biodegradable organic matter which is amenable to biological decomposition within a short hydraulic retention time (Reddy and Graetz, 1988). In addition to this, the media and macrophytes roots in SSF provide a greater number of small surfaces, pores, and crevices where treatment can occur. Moreover, the availability of vast number of organic matter utilizing microorganisms adapted to the aerobic and anaerobic environment of wetland ecosystems (USEPA, 1993; Michael, 2000) make the organic matter removal process more effective.

Microbial degradation and mineralization are the major pathways for BOD_5 and COD removal in constructed wetland system that will result an oxidized byproducts such as CO_2, NO_3^-, SO_4^{2-}, PO_4^{3-} and microbial biomass to the system. In constructed wetland systems, organic matter can also be degraded when taken up by wetland plants (Rencee,

32

2001). Chemical treatments also takes place as certain waste particles contact and react with the media

In addition to BOD_5 and COD, the removal of TSS is used as a wastewater quality parameter to monitor the effectiveness of constructed wetlands for the removal of organic matter. Suspended solids in domestic wastewater include a range of organic and inorganic materials but are typically dominated with by fecal organic matter and organic particles like bacteria. The result showed that TSS removal efficiency of JWBO CW was also high (85%) as that of BOD and COD.

The high removal efficiency of TSS might be due to the fact that, in SSF constructed wetlands the water flows below the ground through gravel and wetland plant roots. This facilitates the physical, chemical and biological wastewater treatment mechanisms such as sedimentation, aggregation, surface adhesion and biodegradation (Mergaert *et al.*, 1992). These characteristics of SSF constructed wetland increase its effectiveness to remove TSS within a short hydraulic retention time, within 2 to 5 days (USEPA, 1993).

JWBO CW effluent concentration values of BOD_5 (2.0 ± 0.6mg/L), COD (68 ± 11.7 mg/L) and TSS (30.0 ± 10.5 mg/L) were compared with the provisional discharge limits values set by National Environmental Quality Standard for domestic wastewater effluent (EEPA, 2003). These limit values were: 80 mg/L for BOD_5, 250 mg/L for COD and 100 mg/L for TSS. The obtained effluent concentration values were by far below the standard limit values, which indicated the effectiveness of constructed wetlands in fulfilling the regulatory limit values to discharge the effluent in to surface and inland water bodies.

6.2 Nitrogen Removal

To analyze the nitrogen removal efficiency of JWBO CW, the influent and effluent NH_4^+-N, NO_3-N and TN were evaluated. Table 2 shows the mean influent and effluent concentrations of each cells and the overall JWBO CW system. The average daily inflow rate of the wetland during the study period was 35,480 liter per day and the influent wastewater's temperature and pH values were 25.8 ± 0.3 ^0C and 7.1 ± 0.27 pH units.

The mean influent parameter values were: ammonium-N (38.7 ± 6 mg/L), nitrate-N (14.7 ± 0.9 mg/L) and total-N (107.7 ± 0.8 mg/L), which was equivalent to the daily loading rate of 2.3 kg /m2/ day, 0.9 kg /m2/ day and 6.5 kg/m2/day, respectively. The influent concentrations were the same, since each cells received equal volume of influent from the equalization tank.

The removal efficiencies of each wetland cells and the overall JWBO wetland were shown in Figure 12. The amount of ammonium-N removed (% removal) in each wetland cells were: cell 1 (24.8%), cell 2 (25.8%), cell 3 (24.8%), cell 4 (23.2%), cell 5 (23.5%) and cell 6 (22.9%). In the same way, the amount of nitrate-N removed by each wetland cell was: 82.5% in cell 1, 78.4% in cell 2, 82.3% in cell 3, 81.0% in cell 4, 77.6% in cell 5 and 81% in cell 6, while for total-N it was 54.5%, 55.7%, 53.4%, 57%, 53.9% and 57%, respectively.

This result showed that wetland cells planted with the same plant species have nearly equal value of removal efficiency, which might be associated with the contribution of different plant species. The overall removal efficiency of JWBO CW was 28.1% for ammonium-N, 64.4% for nitrate-N and 61.5% for total-N, with the corresponding mean effluent concentrations of 27.8 ± 2.1 mg/L, 5.2 ± 0.6 mg/L and 41.5 ± 4.4 mg/L, respectively (Table 2 and Figure 12).

Table 2: The mean ammonium nitrogen, nitrate nitrogen and total nitrogen influent and effluent concentrations (in mg/L) of JWBO CW

Wetland cells	NH$_4^+$-N			NO$_3^-$-N			TN		
	Influent	Effluent	%	Influent	Effluent	%	Influent	Effluent	%
Cell 1[a]	38.7± 6	29.1± 1.8	24.8	14.7 ± 0.9	2.6± .03	82.5	107.7±0.8	49 ± 4.8	54.5
Cell 2[b]	38.7 ±6	28.7± 1.6	25.8	14.7 ± 0.9	3.2± 0.5	78.4	107.7±0.8	47.7 ±4.7	55.7
Cell 3[a]	38.7 ±6	29.1± 1.6	24.8	14.7 ± 0.9	2.6± 0.4	82.3	07.7 ± 0.8	50.5 ±5.5	53.4
Cell 4[c]	38.7 ±6	29.7± 1.6	23.2	14.7 ± 0.9	2.8± 0.6	81	107.7±0.8	44.2 ±5.5	57
Cell 5[b]	38.7 ±6	29.6± 1.7	23.5	14.7 ± 0.9	3.3 ±0.4	77.6	107.7±0.8	49.1 ±4.8	53.9
Cell 6[c]	38.7±6	29.8± 1.8	22.9	14.7 ± 0.9	2.8 ±0.6	81	107.7±0.8	46.3 ±4.0	57
Overall wetland system	38.7±6	27.8± 2.1	28.1	14.7 ± 0.9	5.2 ±0.6	64.4	107.7 0.8	41.5 ±4.4	61.5

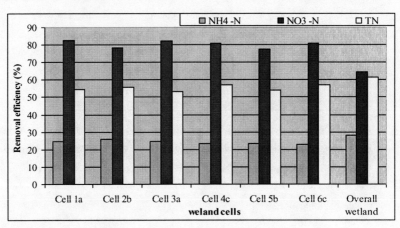

Figure 12: Ammonium N, Nitrate N and Total N Removal Efficiencies of JWBO CW

---

[a] = wetland cells planted with C. papyrus species
[b] =wetland cells planted with C. alternifolia species
[c] =wetland cell planted with P. canariensis

35

Almost similar result values were found in studies done by Pucci *et al.* (2000) nitrate-N (55%), ammonium-N (30%), and total-N (50%); and Kaseva (2003) who obtained with the range of ammonium-N (11.2% - 25.2%) and nitrate-N (32.2% - 44.3%) using CW with the average HRT of 1.93 days.

On the other hand, the highest removal efficiency (92% for NH_4-N and 90% for TN) was obtained in Kenya by Nyakango and VanBurggen (1999). This difference might be due to the design difference; which consists of a combination of SF system followed by SSF wetland cells in serious adjacent to it; this maximizes the retention time and facilitates nitrification process by creating aerobic condition (USEPA, 1993). Moreover, each cells of the wetland were planted with mixed plant species, which maximizes root biomass in the wetland substrate that in turn results aerobic treatments in the root zone (Brix, 1994; Oketch, 2003).

Total nitrogen typically consists of varying proportion of particulate organic nitrogen, dissolved organic nitrogen, ammonium nitrogen, nitrite nitrogen and nitrate nitrogen (Reddy and Patrick, 1984; Kadlec and Knight, 1996). In subsurface constructed wetland system mineralization transforms these organic nitrogen to its inorganic constitutes. Because of the optimum pH value (7.18 ± 0.013), in JWBO CW, the hydrolysis of organic nitrogen resulted mainly ammonium-N. This pathway occurs under both aerobic and anaerobic conditions and it is often referred to as ammonification. Due to this transformation processes, JWBO CW removed 61.5% of the total nitrogen.

Once the ammonium nitrogen is formed and/or entered to subsurface flow wetland system, it can take several possible pathways (Figure 3). The first pathway is nitrification: the aerobic oxidation of ammonium to nitrite by ammonium oxidizing bacteria and the subsequent oxidation of the produced nitrite to nitrate by nitrite oxidizing bacteria. In subsurface flow constructed wetlands nitrification can occur in the oxygenated zones within the rhizosphere of plant roots. While this is a very important process in SSF wetland it is likely that the slow diffusion rate of ammonium from anaerobic zone to root zone (aerobic zone) of the wetland limits the importance of this pathway.

36

The second ammonium pathway is biological uptake. Unlike most terrestrial plants, many aquatic plants use ammonium as a nitrogen source. The removal efficiency of wetland cells planted with the same plant species was nearly similar, which indicates the contribution of different plant species in removing nitrogenous wastes. The result showed that ammonium nitrogen removal efficiency of JWBO CW was very low (28.1%).

This low removal efficiency might be due to the low HRT of JWBO CW (2.16 days). But different studies (USEPA, 1993; Kenneth, 2000; Michael, 2002) indicated that for effective removal of nitrogen the HRT should not be less than 7 days. This long HRT is important because for nBOD to be reduced, cBOD first must be reduced to a relatively low concentration (<40mg/L) that helps to insure adequate degradation of those soluble and simplistic forms of cBOD that inhibits the activity of nitrifying bacteria to occur. Therefore, nitrifying bacteria are dependant on organothrophs to reduce cBOD to obtain energy for cellular activity and carbon for cell synthesis. The other possible reason is the slow diffusion rate of ammonium from anaerobic to aerobic zone in the wetland system.

On the other hand nitrate nitrogen removal efficiency of JWBO CW system was: 82.5% in cell 1, 78.4% in cell 2, 82.3% in cell 3, 81% in cell 4, 77.6% in cell 5 and 81% in cell 6, which was very high than the other forms of nitrogen. In constructed wetland system nitrate is removed using several pathways. The first pathway is denitrification, which usually accounts for the bulk of inorganic nitrogen removal in wetlands. This is because nitrate diffusion rate in wetland soils/substrate are seven times faster than ammonium diffusion rate (Kanlec and Knight, 1996; Simi and Mitchell, 1999; Brix *et al.*, 2003).

Denitrification is the reduction of oxidized nitrogen compounds like nitrate or nitrite to gaseous nitrogen compounds by various chemoorganotrophic and lithoauthotrophic bacteria, which utilize nitrate or nitrite as respiratory electron acceptor to carryout the oxidation of carbonaceous organic matter under anoxic condition (Gersberg *et al.*, 1984; USEPA, 1998; James and William, 1993; Aimee *et al.*, 2000). The second pathway is assimilatory nitrate reduction: when nitrate is taken up and converted to nitrite and then to ammonium by aquatic plants and microorganisms for cell growth (Techobanoglous *et al.*, 2003).

37

The result showed that there was a significant variation between wetland cells and the overall wetland system removal efficiencies in total nitrogen, ammonium nitrogen and nitrate nitrogen. This might be due the aeration effect through the manholes and the ditch. This was because for the overall wetland system removal efficiency analysis, samples were collected after the manholes of each wetland cell and at the end of the ditch (Figure 9)

JWBO CW effluent concentration values of ammonium nitrogen (27.8 ± 2.1 mg/L), nitrate nitrogen (5.2 ± 0.6 mg/L) and total nitrogen (41.5 ± 4.4 mg/L) were compared with provisional discharge limits values set by National Environmental Quality Standard for domestic wastewater effluent (EEPA, 2003). The discharge limit values were: 5 mg/L for ammonium, 20 mg/L for nitrate and 60 mg/L for total nitrogen. Except ammonium nitrogen, the other nitrogenous waste parameter values were below the standard. This showed that with minimum effort to treat ammonium, the overall wetland effluent concentrations met admissible standards set by EEPA to discharge it in to surface and inland water bodies.

This excess ammonium can be removed either by arranging nitrification pre-treatment before the wetland or by treating the effluent with lime or chlorine. Liming converts ammonium ion to ammonia which can be removed from the solution by air stripping. The other treatment option; breakpoint chlorination (supper chlorination), oxidizes ammonium to nitrogen gas.

6.3 Phosphorus Removal

Table 3 presents the results of mean influent and effluent concentrations of each wetland cells and the overall removal efficiencies of JWBO wetland system for orthophosphate (PO_4^{3-}) and total phosphorus (TP).

The mean influent concentration was 8.04 ± 0.8 mg /L for orthophosphate and 8.9 ± 0.95 mg/L for total phosphorus. Based on the average daily inflow of the study period (35,480 liter per day), the average daily loading rate of orthophosphate and total phosphorus to JWBO wetland was 0.5 kg/ m^2 /day and 0.54 kg/ m^2/day, respectively. The influent values were the same for all wetland cells since the volumetric loading rate to each cell was similar.

Figure 13 shows the removal efficiency of each wetland cells and the overall wetland system. The removal efficiency of each wetland cells for orthophosphate was: 24.1% (cell 1), 15.7% (cell 2), 21.6% (cell 3), 23.4% (cell 4), 16.7% (cell 5) and 23.4% (cell 6), while for total phosphorus it was 16.9%, 11.2%, 16.1%, 20.2%, 13.9%, and 16.1%, respectively

.

The overall removal efficiency of JWBO CW was 28% for orthophosphate and 22.7% for total phosphorus, with the final effluent concentration of 5.8 ± 1.1 mg/L and 6.9 ± 1.2 mg/L, respectively. The analysis of variance test result (Annex II: Table 1) showed that the mean effluent concentrations was not significantly ($P > 0.05$) different from the influent concentrations of the wetland, which shows its poor removal efficiency. This finding was in accordance with the findings of Pucci *et al.*, (2000) who obtained 20% TP removal efficiency using subsurface constructed wetland

Table 3: The mean Orthophosphate & total phosphorus influent and effluent concentrations values (mg/L) of JWBO CW

Wetland cells	Orthophosphate			Total Phosphorus		
	Influent	Effluent	% removal	influent	Effluent	% removal
Cell 1[a]	8.04 ± 0.8	6.4 ± 1.4	24.1	8.9 ± 0.95	7.4 ± 1.3	16.9
Cell 2[b]	8.04 ± 0.8	6.8 ± 1.0	15.7	8.9 ± 0.95	7.9 ± 0.96	11.2
Cell 3[a]	8.04 ± 0.8	6.3 ± 1.4	21.6	8.9 ± 0.95	7.5 ± 1.3	16.1
Cell 4[c]	8.04 ± 0.8	6.2 ±1.2	23.4	8.9 ± 0.95	7.1 ± 0.96	20.2
Cell 5[b]	8.04 ± 0.8	6.7 ± 1.2	16.7	8.9 ± 0.95	7.7 ± 0.95	13.9
Cell 6[c]	8.04 ± 0.8	6.2 ±1.5	23.4	8.9 ± 0.95	7.5 ± 1.5	16.1
Overall wetland performance	8.04 ± 0.8	5.8 ± 1.1	28	8.9 ± 0.95	6.9 ± 1.2	22.7

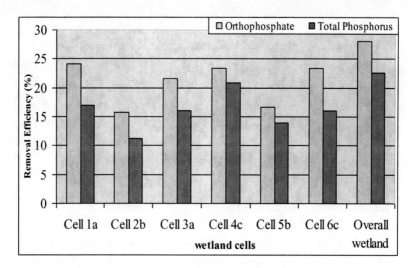

Figure 13: Orthophosphate and Total Phosphorus Removal Efficiencies of JWBO CW

---

[a] =wetland cells planted with C. papyrus species
[b] =wetland cells planted with C. alternifolia species
[c] =wetland cells planted with P. canariensis species

The limiting factor for this low phosphorus removal in this wetland system might be the short hydraulic retention time of the wetland, which was 2.16 days for each cells working in parallel. But according to Miriam *et al.*, (2002) the minimum HRT to remove 50% of the bio-reactive phosphate was 7 days. For example Nyakango and VanBruggen, (1999) obtained 88% removal efficiency for orthophosphate; this difference might be due to high HRT obtained by using serious cells to maximize the contact area and time

The major pathways (Figure 4) that govern the removal of phosphorus in this wetland systems might be through plant assimilation, substrate adsorption, and precipitation reaction which occur when the inflowing water comes in contact with available aluminum ion, calcium, and other clay minerals in the sediment (Faithful, 1996; Kadlec and Knight, 1996).The mineralization of organic matter results in the release of orthophosphate ion to the wetland system. This orthophosphate can then undergo a variety of subsequent reactions.

Once released, orthophosphates is often rapidly taken up in a range of biological growth reactions. The most important of these is the growth and development of biofilms. Overtime these materials is subsequently degraded and recycled in the sediment and incorporated in to wetland macrophytes biomass which has longer storage time. Another very rapid orthophosphate removal pathway which is probably competitive with biological uptake is adsorption of orthophosphate onto available aluminum ion, calcium and other clay minerals (Sinclair, 2000).

JWBO CW effluent concentration values of orthophosphate (5.8 ± 1.1mg/L) and total phosphorus (6.9 ± 1.2 mg/L) were compared with the provisional discharge limits values (5.0 mg/L for orthophosphate and 10.0 mg/L for total phosphorus) set by National Environmental Quality Standard for domestic wastewater effluent (EEPA, 2003). The obtained effluent concentration values showed that the concentration of orthophosphate was slightly higher than the limit value.

41

Since phosphorus is a conservative material, it is extremely important its discharge to the environment is well controlled, because once the system is polluted with phosphorus it can be recycled in a system and result in periods of eutrophication over many years (ponnamperuma, 1972; Flaig and Reddy, 1990; Aisling and Marinus, 2006).

6.4   Sulfate and Sulfide Removal

Table 4 presents the mean sulfate and sulfide concentrations of the influent and effluent concentrations of each wetland cells and the overall wetland systems of JWBO. The average influent concentration was 153.3 ± 17.6 mg/L for sulfate and 4.6 ± 0.5 mg/L, for sulfide. Based on the inflow data (35,480 liter per day), the mean daily loading rate of sulfate and sulfide to JWBO wetland was 9.3 kg/m^2/day and 0.3 kg/m^2/day, respectively. The influent concentration was the same for each wetland cells, since they are supplied with equal volume of wastewater from the equalization tank.

Figure 14 shows the removal efficiency of each wetland cells and the overall wetland of JWBO obtained during the study period. The removal efficiency of each wetland cells for sulfate was: 74.1% (cell 1), 82.8% (cell 2), 73.3% (cell 3), 77.4% (cell 4), 81.2% (cell 5) and 76.7% (cell 6) while for sulfide it was 98.7%, 98.6%, 98.7%, 98.8%, 98.8% and 98.4%, respectively.

The overall removal efficiency of JWBO CW was 77.3% for sulfate and 99% for total phosphorus, with the final effluent concentration of 34.3 ± 0.9 mg/L and 0.047 ± 0.019 mg/L, respectively.

43

Table 4: The mean sulfate and sulfide influent and effluent concentration (mg/L) of JWBO CW

Wetland cells	Sulfate (SO_4^{2-})			Sulfide (S^{2-})		
	Influent	Effluent	% removal	Influent	Effluent	% removal
Cell 1[a]	153.3 ± 17.6	39.7 ± 8.7	74.1	4.6 ± 0.5	0.058 ± 0.019	98.7
Cell 2[b]	153.3 ± 17.6	26.3 ± 1.5	82.8	4.6 ± 0.5	0.064 ± 0.016	98.6
Cell 3[a]	153.3 ± 17.6	41 ± 8.5	73.3	4.6 ± 0.5	0.06 ± 0.017	98.7
Cell 4[c]	153.3 ± 17.6	34.7 ± 1.5	77.4	4.6 ± 0.5	0.056 ± 0.020	98.8
Cell 5[b]	153.3 ± 17.6	28.3 ± 3.2	81.2	4.6 ± 0.5	0.059 ± 0.011	98.7
Cell 6[c]	153.3 ± 17.6	35.7 ± 2.3	76.7	4.6 ± 0.5	0.074 ± 0.013	98.4
Overall wetland performance	153.3 ± 17.6	34.3 ± 0.9	77.3	4.6 ± 0.5	0.047 ± 0.019	99

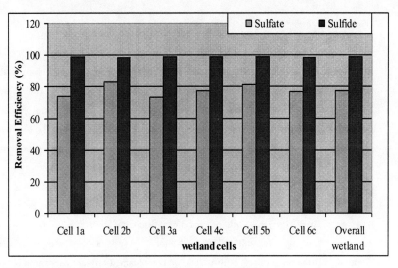

Figure 14: Sulfate and Sulfide Removal Efficiencies of JWBO CW

---

[a] =wetland cells planted with C. papyrus species
[b] =wetland cells planted with C. alternifolia species
[c] =wetland cells planted with P. canariensis species

Microorganisms found in constructed wetland systems preferably utilize electron acceptors that provide the highest energy yield. Oxygen provides the highest energy yield and will be utilized first. Once oxygen is depleted nitrate will be utilized as electron acceptor, if nitrate is depleted sulfate is the next in the sequence of electron acceptors. Therefore, this high removal efficiency of sulfate (77.3%) and sulfide (99%) in JWBO CW system might be due to the net anaerobic environment of the system.

This is because sulfate reduction can take place when sulfate reducing bacteria, which are obligate anaerobes (such as Desulfavibrio bacteria) utilize sulfate as terminal electron acceptor in anaerobic respiration (Aisling and Marinus, 2006). This reaction occurred as the microorganisms assimilate sulfate in the absence of oxygen or nitrate, thus reducing it to sulfide through the transfer of electrons produced by the simultaneous oxidation of the organic compounds (Hsu, 1998; Aisling and Marinus, 2006) (Figure 5). This results the formation of hydrogen sulfide in the system (Rence, 2001).

In constructed wetland system, sulfide is very unstable and readily reacts with free sorbet metal cations forming metal sulfides such as Zinc Sulfide (ZnS), Lead Sulfide (PbS), and Iron Sulfide (FeS). Additionally, it can also react with hydrogen, forming hydrogen sulfide and this will be evolved to the atmosphere as a gas (Morse *et al.*, 1987). The effluent of JWBO constructed wetland has odor problem (the smell of rotten eggs) at the final disposal site, and this might be due to hydrogen sulfide produced through the above stated mechanisms in this wetland system. This unstable nature of sulfide makes the removal efficiency of JWBO CW very effective, which removes 99% of sulfide ion.

JWBO CW effluent concentration values of sulfate (34.3 ± 0.9mg/L) and sulfide (0.047 ± 0.019 mg/L) were compared with the provisional discharge limits values (1000.0 mg/L for sulfate and 1.0 mg/L for sulfide) set by National Environmental Quality Standard for domestic wastewater effluent (EEPA, 2003). The obtained effluent concentration values were by far below the standard values, which showed the effectiveness of constructed wetlands in fulfilling the regulatory limit values to discharge the effluent in to surface and inland water bodies.

6.5   Coliform Bacteria Removal

Table 5 shows the influent and effluent coliform concentrations of each wetland cells and the overall wetland of JWBO. The influent concentration was $6.0 \times 10^7 \pm 6.5 \times 10^5$ for TC and $4.9 \times 10^6 \pm 1.8 \times 10^6$ for FC. The influent concentrations were same, since each cells received equal volume of wastewater in the equalization tank.

Figure 16 shows the removal efficiency of each wetland cells and the overall wetland system. Total Coliform removal efficiency of each wetland cells was: 94.2% (cell 1), 93.8% (cell 2), 94% (cell 3), 94.5% (cell 4), 93.8% (cell 5) and 94.8% (cell 6) while for FC it was 90.2%, 90.4%, 91%, 90.8%, 91% and 91.2%, respectively.

Based on the two month measurements performed from May 14 to July 14, 2007 the overall mean reduction in total and fecal coliforms were 94.5% and 93.1%, respectively with the mean effluent concentration values of $3.3 \times 10^6 \pm 1.1 \times 10^6$ TC and $3.4 \times 10^5 \pm 1.1 \times 10^5$ FC/100ml (Table 5 and Figure 15).

The result of this finding was also similar with the findings of Pucci *et al.* (2000), who reported 99% TC and 99.7% FC removal. ; Kathleen (2000) and Maria *et al.* (2001), each of them reported >90% TC; Kaseva (2003) who obtained 43% - 72% TC and FC removal efficiency of coliform bacteria with constructed wetlands in their study areas.

46

Table 5: **Changes in TC and FC density (cfu/100ml)   at JWBO CW**

Wetland cells	Total Coliform			Fecal Coliform		
	Influent	Effluent	% removal	Influent	Effluent	% removal
Cell 1[a]	6.0×10^7 $\pm 6.5 \times 10^5$	3.5×10^6 $\pm 3.8 \times 10^5$	94.2	4.9×10^6 $\pm 1.8 \times 10^6$	4.8×10^5 $\pm 1.2 \times 10^5$	90.2
Cell 2[b]	6.0×10^7 $\pm 6.5 \times 10^5$	3.7×10^6 $\pm 4.5 \times 10^5$	93.8	4.9×10^6 $\pm 1.8 \times 10^6$	4.7×10^5 $\pm 6.6 \times 10^4$	90.4
Cell 3[a]	6.0×10^7 $\pm 6.5 \times 10^5$	3.6×10^6 $\pm 6.3 \times 10^5$	94,0	4.9×10^6 $\pm 1.8 \times 10^6$	4.9×10^5 $\pm 1.7 \times 10^5$	90
Cell 4[c]	6.0×10^7 $\pm 6.5 \times 10^5$	3.3×10^6 $\pm 5.1 \times 10^5$	94.5	4.9×10^6 $\pm 1.8 \times 10^6$	4.5×10^5 $\pm 1.3 \times 10^5$	90.8
Cell 5[b]	6.0×10^7 $\pm 6.5 \times 10^5$	3.7×10^6 $\pm 4.1 \times 10^5$	93.8	4.9×10^6 $\pm 1.8 \times 10^6$	4.4×10^5 $\pm 8.8 \times 10^3$	91.0
Cell 6[c]	6.0×10^7 $\pm 6.5 \times 10^5$	3.1×10^6 $\pm 3.2 \times 10^5$	94.8	4.9×10^6 $\pm 1.8 \times 10^6$	4.3×10^5 $\pm 3.1 \times 10^4$	91.2
Overall wetland performance	6.0×10^7 $\pm 6.5 \times 10^5$	3.3×10^6 $\pm 1.1 \times 10^6$	94.5	4.9×10^6 $\pm 1.8 \times 10^6$	3.4×10^5 $\pm 1.1 \times 10^5$	93.1

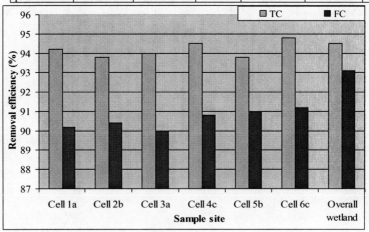

Figure 15:   Total Coliform and Fecal Coliform Removal Efficiency of JWBO CW

[a] =wetland cells planted with C. papyrus species
[b] =wetland cells planted with C. alternifolia species
[c] =wetland cells planted with P. canariensis species

47

This decline in coliform bacteria concentration could be attributed to biotic and abiotic factors. The removal of coliform bacteria in wetlands is essentially a two stage process. Most microorganisms including coliform bacteria are organic particulates (Suresh and Bruce, 2003). The initial stage of coliform bacterial removal is therefore particle removal. This occurs via the same process in suspended solids: sedimentation, surface adhesion and aggregation (Rogers, 1983; Joseph, 2005). A serious of other processes that may occur both before and after coliform bacteria particles have been removed from the water column are important in influencing the viability of coliform bacteria.

The major processes in this category are: the hostility of the environmental conditions (temperature, salinity, and turbidity, reduction of organic matter content), predation and infection by other microbes (Shiaris, 1985; Faithful, 1996). The result of this study indicated that there was a statistically significant (p< 0.05), at the 0.01 level, correlation between coliform bacteria reduction with wastewater temperature and BOD_5 values of the final effluent with the calculated R value of -0.471 (R^2 = 0.222) and -0.683 (R^2 = 0.467), respectively. The presence of wetland plant and gravel not only improves coliform trapping but also creates a hostile environment.

In subsurface flow constructed wetland the presence of wetland plants and gravel media can play a crucial role in increasing the efficiency and effectiveness of coliform bacteria removal. They improve the trapping efficiency for small particles by increasing the surface area of biofilms in the flow path. Small particles are trapped by surface adhesion on the biofilms (Gresberg *et al.*, 1989). Similarly the efficiency of coliform bacteria removal by sedimentation or enhanced sedimentation within planted subsurface areas is improved. The root zone of wetland plant is a highly active biological community supporting a range of metabolic processes

The performance efficiency results indicated that this wetland system has excellent pathogen removal capability, though the mean final effluent concentration of TC was above the National Effluent Emission Standard limit values (400 cfu/100ml) set by Environmental Protection Authority of Ethiopia. But as shown in Figure 16, the overall final effluent of JWBO wetland is disposed in the nearby stream in which the local people use it for different purposes; such as for body washing, vegetable production, animal fattening, recreation.

The presence of coliform bacteria, (which in turn indicates the presence of pathogenic microorganisms) in wastewater effluent above the emission standard makes the receiving water unsuitable for direct contact recreational use and some times unsuitable for use as source of water for a public supply (Christone, 2004).

Direct using of the effluent for hand and face washing	Fetching the stream for household use and animal fattening

Figure 16: Pictures which show when local people using the effluent as well as the stream receiving the effluent from the treatment plant

However, one strong advantage of using constructed wetlands to treat wastewater over natural wetlands is that the final effluent can be easily chlorinated (Rence, 2001). In addition to fulfilling the National Emission Standards of the country, chlorine disinfection of constructed wetland effluent can produce waters suitable for unrestricted use (USEPA, 1998). For example a study conducted in Australia (Sinclair, 2000) showed that 30% of the constructed wetland in the country uses the effluent for irrigation of Golf courses, woodlots and parks.

Therefore, JWBO should chlorinate its effluent to fulfill the provisional discharge limits values set by National Environmental Quality Standard for domestic wastewater effluent as well as to recycle the wastewater to use it for different purposes in their campus. Additionally, they can supply to the nearby community, which are currently suffering due to shortage of water during dry seasons.

6.6 The contribution of different wetland plant species for domestic Wastewater treatment

The treatment efficiencies of JWBO wetland cells covered with *Cyprus papyrus*, *Cyprus alternifolia* and *Phoenix canariensis* plant species were investigated throughout the study period. The changes in the mean influent and effluent concentrations values and the removal efficiency of the measured physicochemical and microbiological variables of the wetland cells are shown in Table 6.

Wetland cells planted with *Cyprus papyrus* (cell 1 and 3) showed higher removal efficiency for: NO_3-N (82.4%), NH_4^+-N (24.8%), TN (54.8%), PO_4^{3-} (23.5%), and TSS (83.9%) than the other wetland cells planted with other plant species. Similarly wetland cells planted with *Phoenix canariensis* (cell 4 and 6) showed higher removal efficiency for: TP (17%), S^{2-} (99%), BOD_5 (98%), COD (90%), TC (94.7%) and FC (91%).

While wetland cells planted with *Cyprus alternifolia* (Cell 2 and 5) showed higher removal efficiency only for SO_4^{2-} (82.2%) than the others (Table 6). The analysis of variance (ANOVA) test showed that the contribution of plant species were statistically significant, ($p < 0.05$) for all selected wastewater parameters (Annex II, Table 2).

Table 6: The mean influent and effluent concentrations (mg/L) along with the mean removal efficiency of wetland cells planted with *C. papyrus, C. alternifolia* and *P. canariensis* plant species at JWBO CW

wastewater parameters	Cyprus papyrus[a]			Cyprus alternifolia[b]			Phoenix canariensis[c]		
	Influent	Effluent	% removal	Influent	Effluent	% removal	Influent	Effluent	% removal
$NO_3^- - N$	14.7	2.6	82.4	14.7	3.2	78	14.7	2.8	81
$NH_4^+ - N$	38.7	29.1	24.8	38.7	29.2	24.7	38.7	29.8	23
TN	107.7	54.8	49.1	107.7	55.4	48.6	107.7	55.8	48
$PO_4^{3-} - P$	8.04	6.2	23.5	8.04	6.8	16.2	8.04	6.2	23
TP	8.9	7.5	16.5	8.9	7.8	13.3	8.9	7.5	17
SO_4^{2-}	153.3	40.4	73.7	153.3	27.3	82.2	153.3	35.2	77
S^{2-}	4.6	0.059	98.7	4.6	0.062	98.7	4.6	0.065	99
TSS	201.3	32.6	83.9	201.3	35.7	82.3	201.3	33.9	83
BOD_5	273.3	5.8	97.9	273.3	6.2	97.8	273.3	5.7	98
COD	619.3	65.17	89.5	619.3	78.33	87.4	619.3	64.17	90
TC, cfu/100ml	6.0×10^7	3.7×10^6	93.9	6.0×10^7	3.9×10^6	93.6	6.0×10^7	3.9×10^6	94
FC, cfu/100ml	4.9×10^6	4.7×10^5	90.4	4.9×10^6	4.7×10^5	90.4	4.9×10^6	4.6×10^5	91

a = cell 1 and 3
b = cell 2 and 5
c = cell 4 and 5

This difference might be mainly because plants help the treatment process of wetlands in several ways such as filter wastes, regulate flow, provide surface area for microbiological treatment, provide shed and control algae growth, contribute oxygen to the cells, take up and store some of metals and nutrients from the wastewater (Sinclair, 2000; Joseph, 2005).

But all wetland plant species didn't have the same capacity to use the above stated mechanisms to remove different pollutants found in the wastewater (Gersberg *et al.*, 1985; Joseph, 2005). For example, study conducted by Kuet *et al.*, (1999) has shown that wetland planted can directly uptake to 20% of the nutrients found within the treatment effluent depending on plant type of the wetland. Another study conducted by Brix, (1994), also showed that the uptake species of emergent macrophytes was 50 to 150 kg phosphorus /ha/year and 1000 to 25,000 kg of nitrogen /ha/year.

The root zone of aquatic plants is also primary site for pollutant uptake and transformation as it is a zone of oxygen transfer between the plant and sediment microbial activity and pollutant oxidation. This also will be depending on the potential root penetration depth and root mat structure of the plant species (USEPA, 1993; Faithful, 1996; Joseph, 2005).

7. CONCLUSION AND RECOMMENDATIONS

7.1 Conclusion

The need for improvement and conservation of the environment in Ethiopia is necessitating the provision of energy and cost effective secondary wastewater treatment facilities for small communities such as schools, hospitals, military camps, colleges, farms, industries, and universities where on-site wastewater disposal technology is predominant.

Constructed wetland system operates using natural processes and usually do not require substantial energy inputs. The biological processes are typically solar-driven as light and carbon sources (from the substrate) are used to derive the microbial and plant processes. Therefore constructed wetlands seem to be appropriate since in Ethiopia there is a year round suitable climatic condition for rapid biological growth, which influence the treatment process in wetlands. But in Ethiopia this technology has not yet been recognized as a treatment option for the wastewater management. Failure for the acceptance is due to the lack of wetland information and research work to prove the effectiveness of the wetland for wastewater treatment.

This result showed that the average percentage removal efficiency of JWBO wetland system was: 99.3% (BOD_5), 89% (COD), 85% (TSS), 28.1% (NH_4^+-N), 64% (NO_3-N), 61.5% (TN), 28% (orthophosphate), 22.7% (TP), 77.3% (Sulfate), 99% (Sulfide), 94.5% (TC) and 93.1% (FC). These showed that the treatment performance of JWBO CW was low for ammonium nitrogen, and phosphorus. This was mainly due to the low HRT of the wetland.

Treatment is necessary to correct wastewater characteristics in such away that the use of final disposal of the treated effluent can take place in accordance the rules set by the relevant legislative bodies without causing adverse impacts on receiving water bodies. The result showed that except ammonium nitrogen, coliform bacteria and orthophosphate, the other studied parameters effluent concentration values were by far below the standard discharge limit values set by National Environmental Quality Standard for domestic wastewater effluent (EEPA, 2003). This showed the effectiveness of constructed wetlands in fulfilling the regulatory limit values to discharge the effluent in to surface and inland water bodies.

The result of this study also showed that wetland cells planted with *Cyprus papyrus* (Cell 1 and 3) showed higher removal efficiency in NO_3-N (82.4%), NH_4^+-N (24.8%), TN (54.8%), PO_4^{3-}-P (23.5%), and TSS (83.9%) than the other wetland cells planted with the other two plant species. Similarly wetland cells planted with *Phoenix canariensis* (cell 4 and 6) showed higher removal efficiency in TP (17%), S^{2-} (99%), $BOD5_5$ (98%), COD (90%), TC (94%) and FC (91%).

While wetland cells planted with *Cyprus alternifolia* (cell 2 and 5) showed higher removal efficiency only for SO_4^{2-} (82.2%) than the other wetland cells. In addition to the treatment performance, the three wetland plant species planted at JWBO wetland helps to increase the aesthetic value of their compass, especially for their Conference Hall constructed near the wetland and their compass in general.

Generally we can conclude that the treatment performance of JWBO wetland system was very encouraging in promoting the use of constructed wetlands as an alternative wastewater treatment system for protecting sensitive water bodies that receive partially treated or untreated effluents. In addition, for developing countries like Ethiopia, that have limited resources for the construction and operation of conventional treatment plants, constructed wetlands are the most economical solutions.

This study indicated the SSF wetlands treatment system can effectively treat wastewaters; fortunately the climatic condition of Ethiopia is favorable for the growth of different wetland plant species which enhance the removal efficiency of constructed wetland system. Therefore, for our country wetland plant species selection and management techniques that create the largest rhizosphere surface area per volume of bed and bed design (optimal depth, HRT, and media) should be explored further.

7.2   Recommendations

Based on the results of this case study and other research outputs done in other countries (with similar conditions); in order to use this technology in our country as alternative wastewater treatment technology and to improve the performance of JWBO CW, the following points are recommended:-

- For effective wastewater treatment performance, constructed wetlands should consist of a minimum of two to three cells in serious and all the cells should be planted with different plant species within the system that will increase the root biomass in the wetland substrate as well as its biodiversity.

- The significant phosphorus and nitrogen removal will require a long detention time in the wetland system. The longer the wastewater remains in the wetland, the greater chance of sedimentation, biotic processing and retention of nitrogen and phosphorus nutrients. Then, if the major objective of the treatment wetland is to remove nitrogenous and phosphorus wastes, then the wetland should be designed with HRT not less than 7 days

- This case study indicated that constructed wetland treatment systems can effectively treat domestic wastewater. But different wetland plant species selection and management techniques that create the largest rhizosphere surface area per volume of bed and bed design (optimal depth, HRT and media type) should be explored with further research.

- The effluent concentration values of ammonium and coliform bacteria were above the discharge limits set by National Environmental Quality Standard for domestic wastewater effluent. Then additional effluent treatment options are required.

- During the assessment done in different wetlands found in the CRV of Ethiopia, and Bahir-Dar it is confirmed that these wetlands have different original pristine plant types. Therefore an inventory and detailed studies of these plant species compositions with their contribution for wastewater treatment needs further research.

- Despite suitable climatic conditions in Ethiopia, until now no efforts have been made to investigate the effectiveness of constructed wetlands in treating various types of wastewaters. So in future a detail research that incorporates all issues of wetland should be done.

- In addition, an advocacy and awareness creation works should be done by concerned institutions in the country on the methods of wastewater treatment that are not only cheap and highly effective like constructed wetlands but also which promotes resource conservation and environmental protection.

- Government regulations and legislations need to be enforced in order to ensure that polluters meet environmental standards of effluent discharge in to water bodies and natural wetlands found in different parts of the country.

8. REFERENCES

1. **AAWSA** (2001). Wastewater Master Plan, Main Report on the Design Study of Addis Ababa Sanitation Services Improvement Project. Addis Ababa, Ethiopia. Pp1-8

2. **AAWSA (2003).** Concept Report on Basic Design Services for Addis Ababa Water and Sewerage Project Component II. pp 3-11

3. **Abebe, Y. and Geheb, K.** (2003). Wetlands of Ethiopia. Proceeding of a Seminar on the Resources and Status of Ethiopia's Wetlands. IUCN Wetlands and Water Resources Program Blue Serious. pp1-11

4. **Abe, K., Ozaki, Y. and Kihou, N.** (1997). Introduction of Fiber Plants to Plant Bed Filter System for Wastewater Treatment in Relation to Resource Recycling. J. Soil Science and Plant Nutrition. pp35-43

5. **Addis Ababa city Municipality, Ministry of Works & Urban Development and GTZ** (2004). (Unpublished) Wastewater Management for the Newly Constructing Houses (Condominiums) of Piaza Projects.

6. **Aalbers, H.** (1998).Resource Recovery from Fecal Sludge Using Constructed Wetlands: A survey of the Literature. UWEP Working Document. The Netherlands (www.waste.nl/redir/content)

7. **Afework, H.** (2006). An Overview of Distribution, Socio-ecological Significance, Management and Threats of Wetland Ecosystems in Ethiopia. A conference proceeding on "Environment for Survival", Taking Stock of Ethiopia's Environment, held in Addis Ababa, 2-4 October 2006. Green Forum, vol. 1. pp99-127

8. **Aimee, M., Parkin, G. and Wallace, S.** (2000). A Comparison of Constructed Wetlands Used to Treat Domestic Wastes; Conventional, Drawdown, and Aerated Systems. Proceedings of the Wetland Systems for Water Pollution Control Conference, International Water Quality Association. 2000. (www.efka.utm.my/thesis/

9. **Aisling, D. and Marinus, L.** (2006). Using Ecosystem Processes in a Constructed Wetland to Treat Mine Wastewater in Irelands. National University of Ireland, at Dublin. In press (Pre-printed with Kind permission) in Encyclopedia of Water. Weley publisher. (www.ucd.ie/wetland)

10. **APHA/American Water Works Association/Water Environment Federation** (1998). Standard Methods for the Examination of Water and Wastewater. 20th edition, Washington, D.C., USA

11. **Baird, C., Rice, J., Rashash, D. and Humenik, F.** (2000). Constructed Wetlands for Swine Wastewater Treatment. Department of Biology and Agricultural Engineering, North Caolina State University. (www.wctin.com)

12. **Brix, H**. (1994). Functions of Macrophytes in Constructed Wetlands. J. Water Science and Technology. Vol. 29. No 4. pp 71-78

13. **Brix, H., Arias, C. and Johansen, N.** (2003). Experiments in a Two Stage Constructed Wetland System; Nitrification Capacity and Effects of Recycling on Nitrogen Removal. (www.euronet.nl/users/backhuys/newsbull-8pdf)

14. **Chris, C.** (1997). Guidelines for Constructed Wetland in New Zealand. J. NIWA Science and Technology Series. No 48

15. **Christina, L.** (2005). Nutrient Removal Using a Constructed Wetland in Southern Quebec. Masters Thesis. Department of Bio-resource Engineering, McGill University, Montréal

16. **ESTA/CPC** (2004). (Unpublished) Proposal for Organizing National Roundtable on the Sustainable Consumption and Production in Akaki River Basin. (www.ics.trieste.it/documents/download/df2990.pff)

17. **EEPA / UNIDO** (2003). Standards for Industrial Pollution Control in Ethiopia. Part Three, Standards for Industrial Effluents (General). FEPE of Ethiopia. Vol. 1. pp45-48

18. **EIBC (2007)**. Ecosystems of Ethiopia. Ethiopian Institute of Biodiversity Conservation. Institute of Biodiversity Conservation. www.gov.et

19. **Franson, M.** (1985). Standard Method for the Examination of Water and Wastewater, American Public Heath Association, Washington, D.C.

20. **Faulkner, S.** (2004). Soils and Sediment: Understanding Wetland Biogeochemistry. Wetlands–Exploring Environmental Challenges: A Multidisciplinary Approach. (www.nwr.usgs.gov)

21. **Faithful, J.** (1996)(Unpublished). The Fate of Phosphorus in Wetland, Review Report for the Queensland Department of Natural Resource. Australian Centre for Tropical Freshwater Research. (www.actfr.juc.edu.av/publications)

22. .**Fauque, G.** (1995). Ecology of Sulfate-Reducing Bacteria, in Constructed Wetland. Plenum Press, New York. (www.springerlink.com)

59

23. **Getahun, W. and Adinew, A.** (1999) Wastewater Management in Addis Ababa. 25th WEDC Conference Proceeding on Integrated Development for Water Supply and Sanitation. www.ihwb.tu.darmstadt.ed.

24. **Gersberg, R., Brenner, R., Lyon, S. and Elkins, B.** (1984). Survival of Bacteria and Viruses in Municipal Wastewaters Applied to Artificial Wetlands. Graduate School of Public Health, Santiago State University. Magnolia publishing. pp 237-245. www.publichealth.sdsu.edu/facultydetail.php

25. **Gersberg, R., Elkins, B. and Goldman, C.** (1983). Use of Artificial Wetland to Remove Nitrogen from Wastewater. J. WPCF. Vol. 56. No. 2. pp152-157

26. **Gersberg, R.,** Elkins, B. and Goldnman, C. Lyon, S. (1985). Role of Aquatic Plants in Wastewater Treatment by Artificial Wetlands. J. Water Resource. Vol.20. No.3. pp363.368

27. **Gearheart, R.** (2004). Planning and Preliminary Technical Study for the Application of a Wetland Treatment System at the City of Fort Bragg wastewater Facility. www.ci.fort-bragg.ca.us

28. **Godfrey, P., Kaynor, E., Pelezarski, S. and Benforado, J.** (1985). Ecological Considerations in Wetlands Treatment of Municipal Wastewaters, Van Nostrand Reinhold Company, New York. pp 474. www.scwrp.org/document/hottopics/

29. **Hammer, D.** (1989). Constructed Wetlands for Wastewater Treatment: Municipal, Industrial and Agricultural. Lewis Publisher, Inc. Michigan, Chelsea.

30. **Healy, M. and Cawley, A.** (2001). Wetland and Aquatic Processes: Nutrient Processing Capacity of a Constructed Wetland in Western Ireland. J. Environmental. Quality. Vol. 31. pp1739 – 1742

31. **Hester, E. and Harrison, M.** (1995). Waste Treatment and Disposal. J. Environmental Science and Technology. Vol. 30. pp35 - 45

32. **Hsu, S.** (1998). The Use of Sulfur Isotopes to Determine the Effectiveness of Sulfate-Reduction in the Remediation of Acid Mine Drainage at Wills Creek Constructed Wetland. Masters Thesis, Dept of Geology, University of Cincinnati, College of Arts and Sciences.

33. **John, I. and Patrick, C.** (2002). Design Analysis of a Recalculating Living Machine for Domestic Wastewater Treatment. Proceeding of the 7th International Conference on

Wetland Systems for Water Pollution Control. International Water Association, Orlando FL. pp 547 – 555.

34. **Joseph, K.** (2005). Optimizing Processes for Biological Nitrogen Removal in Nakivubo Wetland, Uganda. Royal Institute of Technology, Department of Biotechnology. Doctoral Thesis, Stockholm, Sweden.

35. **Kathleen, M.** (2000). Analysis of Residential Subsurface Flow Constructed Wetland Performance in Northern Alabama. J. Small Flows. Vol. 1. No. 3. PP 34– 39

36. **Kaseva, M.** (2003). Performance of a Subsurface Flow Constructed Wetland in Polishing Pre-treated Wastewater, A tropical Case Study. J. Water Research. Vol. 38.pp681-687. www.elsevier.com/locate/waters

37. **Knight, R.** (1990). Wetland systems; Natural Systems for Wastewater Treatment. Manual of Practice FD-16. J .of Alexandrian Water Pollution Control Federation. pp211- 260.

38. **Kenneth, D.** (2001). Using Constructed Wetlands for Removing Contaminants from Livestock Wastewater. Ohio State University Fact Sheet, School of Natural Resources, Columbus, Ohio.

39. **Kyambadde, J., Kansiime, F., Gumaelius, L. and Dalhammar, G.** (2004). A Comparative study of *Cyprus papyrus* and *Miscanthidium violaceum*-based constructed wetlands for wastewater treatment in a tropical Climate. J.ELSEVIER Water Research. No. 38. Pp475-485.

40. **Luise, D.,Robert, E., Lamonte, G., Barry, I., Jeffrey, L., Timonthy, B., Glenn, R., Melanie, S., Charles, T. and Harold, W.** (1999). The Handbook of Constructed Wetlands. A Guide to Creating Wetlands for: Agricultural Wastewater, Domestic Wastewater, Coal Mine Drainage and Storm Water in the Mid-Atlantic Region. General Considerations. Vol. 1

41. **Luckeydoo, L., and Fausey, B.** (2002). Early Development of Vascular Vegetation of Constructed Wetland in Northwest Ohio Receiving Agricultural Waters. J. Agricultural Ecosystems and Environment. pp 88 - 94

42. **Indian Natural Resources Conservation Service** (2001). Conservation Practice Standard for Constructed Wetlands, code 658. www.sci.journals.org/cgi/content/ful

43. **MacDonald, L.** (1994). **Water** Pollution Solution: Build a Marsh. J. American Forests. Vol. 100. No.7/8, pp 26-30

44. **Mariade'j, Q., Martine, M., Eric, D. and Charles, P.** (2001). Removal of Pathogenic and Indicator Microorganisms by a Constructed Wetland Receiving Untreated Domestic Wastewater. J. Environmental Science and Health, Part A: Vol. 36. No.7.

45. **Muhammad, M., Baiy, M., Murtatu, M., and Ishtiaq R.** (2004). Constructed Wetland: An Option for Wastewater Treatment in Pakistan. EJ. Environmental, Agricultural and Food Chemistry. pp 739– 742

46. **Mashauri, D., Mulungu, D., and Abdulhussein, B.** (2000). Constructed Wetland at the University of Dares Salaam. J. Water Research. Vol. 34. No 4, pp1135-1144

47. **Martha, S.** (2003). Habitant Value of Natural and Constructed Wetlands Used to Treat Urban Runoff. Literature Review. A Report Prepared for the California State Coastal Conservancy. Southern California Coastal Water Research Project. (www.sccwrp.org)

48. **Morse, J., Millero, F., Cornwell, J., and Richard, D.,** (1987). The Chemistry of the Hydrogen Sulfide and Iron Sulfide Systems in Natural Waters. J. Earth Science Reviews. No. 24, pp1- 42. www.canterbury.ac.nz

49. **Michael, H.** (2002). Wastewater Microbiology: Nitrification and Denitrification in the Activated Sludge Process. John Wiley & Sons Publisher. www.amazon.com

50. **Newman, J., Clausen, J. and Neafsey, J.** (2000). Seasonal Performance of a Wetland Constructed to Process Diary Milk house Wastewater in Connecticut. J. Ecological Engineering. Vol. 14. pp181-198

51. **Nichols, A.** (1983). Capacity of Natural Wetlands to Remove Nutrients from Wastewater. J. Alexandrian Water Pollution Control Federation. pp495 - 505

52. **Njanu, K. and Mlay, H** (2000). Wastewater Treatment and other Research Initiatives with Vetiver Grass. University of Dar e salaam, Prospective College of Engineering and Technology, Department of Chemical and Process Engineering.

53. **Nzengy'a, D., and Wishitemi, B.** (2001). The Performance of Constructed Wetlands for Wastewater Treatment: A Case Study of Splash Wetland in Nairobi, Kenya. J. Hydrological Process. Vol. 15. pp3239-3247

54. **Nyakango, J. and VanBruggen, J.** (1999). Combination of a well Functioning Constructed Wetland with a Pleasing Landscape Design in Nairobi, Kenya. J. Water Science and Technology. Vol. 40. No. 3. pp247-247

55. **Okurut, T., Rijs, G., and VanBruggen, J.** (1999). Design and Performance of Experimental Constructed Wetlands in Uganda, Planted with *Cyprus papyrus* and *phragmites mauritianus*. J. Water Science and Technology. Vol. 40.No. 3. pp265-271

56. **Oketch, M.** (2003). The Potential Role of Constructed Wetlands in Protection and Sustainable Management of Lake Catchments in Kenya. Department of Environmental Science, Egerton University, Kenya. www.iodewebl.vliz.be/bitstream/

57. **Pottir, C. and Korathonasis, A.** (2001). Vegetation Effects on the Performance of Constructed Wetlands Treating Domestic Wastewater. Proceeding of the 9th National Symposium on On-site Wastewater Treatment (11-14March, 2001 Forth Worth, Texas, USA). American Society of Agriculture and Biological Engineers. pp 662 – 672.

58. **Pucci, B., Conte, G., Martinuzzi, N., Giovannelli, L. and Masi, F.** (2000). Design and Performance of a Horizontal Flow Constructed Wetland for Treatment of Diary and Agricultural Wastewater in the Chianti Countryside. www.provincia.pistoia.it

59. **Ramsar Convention Bureau (1997).** Economic Evaluation of Wetlands. A Guide for Policy Makers and Planners. The University of York, Institute of Hydrology. Gland, Switzerland. pp 1- 55

60. **Renee, L.** (2001). Constructed Wetlands: Passive System for Wastewater Treatment. Technology Status Report, prepared for the USEPA Technology Innovation office under a National Network of Environmental Management Studies Fellowship.

61. **Reddy, K.** (2004). Phosphorus Cycling in Wetlands Associated with Agricultural Water Shades. 'Nutrient Management in Agricultural Watersheds–A wetland Solution-'A symposium Conducted at Teargases Research Center, Johnstown Castle, Cowexfored, Ireland 24-25 May,2004

62. **Richard, E.** (1998). Demonstration/Evaluation of Constructed Wetlands as an Alternative On-site Wastewater Treatment System. Texas Agricultural Extension Service's Project, Department of Agricultural Engineering. Galveston Bay Information Center. www.gbic.tamug.edu

63. **Rogers, K., Breen, P. and Chick, A.** (1991). Nitrogen Removal in Experimental Wetland Treatment Systems. Evidence for the Role of Aquatic Plants. J. Archives of Environmental Contamination and Toxicology. Vol. 41. No. 3. Pp274-281.

64. **Rocky Mountain Institute (1998).** Green Development: Integrating Ecology and Real Estate. John Wiley and Sons. Inc. Toronto, Ontario. Pp 146-150.

65. **Sebsebe, D., Sue, E., and Inga, H.** (1997). Flora of Ethiopia and Eritrea, Hydrocharitaceae to Arecaceae. Addis Ababa, Ethiopia. Vol. 6. pp 434-435, 441-442 and 518.

66. **Smith, R.** (1999). Lectures in Wastewater Analysis and Interpretation. Genuine Publishing Corporation. www.genium.com

67. **Georgia Environmental Protection Division, Water Protection Branch, Engineering & Technical Support Program** (2002). Guidelines for Constructed Wetland for Municipal Wastewater Facilities. www.gaepd.org

68. **Sinclair, K.** (2000). Guidelines for Using Free Water Surface Constructed Wetlands to Treat Municipal Sewage. Queensland Department of Natural Resources.

69. **Simi, A. and Mitchell, C.** (1999). Design and Hydraulics Performance of a Constructed Wetland Treating Oil Refinery Wastewater. J. Water Science and Technology. Vol. 40. No 3. pp 301-307

70. **Suresh, D. and Bruce, L.** (2003). Removal and Fate of Specific Microbial Pathogens within On-site Wastewater Treatment Systems. A Report Prepared for Texas on-site Wastewater Treatment Research Council, Texas. www.wrri.msstate.edu/se-toc/

71. **Tanner, C. and Sukias, J.** (2003). Linking Pond and Wetland Treatment: Performance of Domestic and Farm Systems in New Zealand. J. Water science and Technology. Vol. 2. No. 48. pp 331-339.

72. **Tchobanoglous, G.** (1997). Ecological Engineering for Wastewater Treatment. 2nd edition. LEWIS Publisher-CRC-TAYLOR & Francis.

73. **Tchobanoglous, G., Burton, F. & Stensel, H.** (2003). Wastewater Engineering: Treatment, Disposal, and Reuse. 4th edition, McGraw-Hill, Inc. Book Company. New York.

74. **TomOkin, O.** (2000). A Pilot Study on Municipal Wastewater Treatment Using a Constructed Wetland in Uganda. UNESCO-IHE Dissertation, Serious Information. J. Taylor & Francis Engineering, Water and Earth Sciences. www.blakema.nl/series/

75. **UNDP** (2006). Wastewater as a controversial; Human Development Report office, Occasional Paper. www.hdr.undp.org/hdr2006/

76. **UNESCO** (1994). Convention on Wetlands of International Importance Especially as Waterfowl Habitat. Ramsar, Iran, 1971 as amended by the Protocol of 3. 12. 1982 and the amendment of 28. 5. 1987. Director, Office of International Standards and Legal

Affaires, United Nations Educational, Scientific and Cultural Organization (UNESCO), Paris.

77. **USEPA** (1988). Design Manual: Constructed Wetlands and Aquatic Plant Systems for Municipal Wastewater Treatment. EPA 625/1-88/022, Office of Research and Development, Washington, D.C.

78. **USEPA** (1993). Subsurface Flow Constructed Wetlands for Wastewater Treatment: A Technology Assessment. EPA 832-R-93-001, Office of Water, Washington, D.C.

79. **Vymazal, J**. (1996). Subsurface Horizontal Flow Constructed Wetland for Wastewater Treatment: The Czech Experience. J. Wetlands Ecology and Management. Vol. 4. No. 3. Pp199-206

80. **Vymazal, J**. (2002). The Use of Subsurface Flow Constructed Wetlands for Wastewater Treatment in the Czech Republic: Ten Years of Experience Report. J. Ecological Engineering. Vol. 18/2002. Pp633-646.

81. **Wallance, S**. (1998). Putting Wetlands to Work. J. American Society of Civil Engineers. J. Civil Engineering. Vol. 68. No. 7. Pp57-59.

82. **William, W.** (1997). Design Features of Constructed Wetlands for Non-point Source Treatment. Indian University, Schools of public and Environmental Affairs, Bloomington, Indiana. www.spea.indiana.edu/clp/

83. **William, J. and James, G** (1993). Wetlands. 2nd edition. Van Nostrand Reinhold Company, New York. pp 3- 24, 114 – 147 and 577 - 592

84. **Zuidervaart, I., Tichy, R., Kvet, J. and Hezina, F.** (1999). Distribution of Toxic Metals in Constructed Wetlands Treating Municipal Wastewater in the Czech Republic. Backhuys publishers, Leiden, the Netherlands.

85. **Zerihun, W. and Kumlachew, Y.** (2003). Wetland Plants in Ethiopia with Examples from Illubabur, South-Western Ethiopia. Proceedings of a Seminar on the Resources and Status of Ethiopia's Wetland. IUCN Wetlands and Water Resources Program Blue Serious. pp49-58

ANNEX I. Mean, Minimum and Maximum influent and effluent concentration values of JWBO CW

Table 1: Mean, Std. Error, Minimum and Maximum influent and effluent nitrate N, ammonium N and total N concentration values of JWBO CW

code of the cell		Nitrate -Nitrogen (No3 -N), mg/L	Ammonium ion (NH4+),mg/L	Total Nitrogen (TN), mg/L	Orthophosphate (PO4-P), mg/L	Total Phosphate (TP), mg/L
Influent	Mean	14.7000	38.6667	107.6733	8.0433	8.8733
	Std. Error of Mean	.85049	6.00481	.81407	.67716	.95016
	Minimum	13.10	30.00	106.42	6.80	7.13
	Maximum	16.00	50.20	109.20	9.13	10.40
cell 1	Mean	2.5667	27.1000	49.0233	5.8033	6.4667
	Std. Error of Mean	.03333	1.72143	4.80793	1.40286	1.31318
	Minimum	2.50	24.60	40.17	4.21	4.60
	Maximum	2.60	30.40	56.70	8.60	9.00
cell 2	Mean	3.1667	28.7000	47.6967	6.7833	7.7800
	Std. Error of Mean	.52387	1.57162	4.69050	.99596	.96423
	Minimum	2.50	25.60	38.60	5.05	6.10
	Maximum	4.20	30.70	54.23	8.50	9.44
cell 3	Mean	2.6333	29.0667	50.4700	6.2667	7.4667
	Std. Error of Mean	.44096	1.58360	5.45594	1.37760	1.29786
	Minimum	1.80	25.90	40.17	4.60	5.00
	Maximum	3.30	30.70	58.74	9.00	9.40
cell 4	Mean	2.8333	29.6667	44.2433	6.1600	7.0933
	Std. Error of Mean	.57831	1.58990	5.19199	1.22513	.95815
	Minimum	1.70	26.50	33.90	4.18	5.60
	Maximum	3.60	31.50	50.21	8.40	8.88
cell 5	Mean	3.3333	29.5667	49.0967	6.7000	7.6600
	Std. Error of Mean	.43716	1.73429	4.78874	1.24231	.95023
	Minimum	2.80	26.10	40.00	4.60	6.10
	Maximum	4.20	31.40	56.24	8.90	9.38
cell 6	Mean	2.8333	29.7667	46.2733	6.1667	7.4867
	Std. Error of Mean	.64893	1.80954	4.02542	1.53768	1.50936
	Minimum	1.60	26.30	38.60	3.90	4.50
	Maximum	3.80	32.40	52.22	9.10	9.36
Final effluent	Mean	5.1667	27.7667	41.4700	5.7867	6.8800
	Std. Error of Mean	.63596	2.06263	.69376	1.12761	1.17717
	Minimum	4.10	23.70	40.17	3.66	4.70
	Maximum	6.30	30.40	42.54	7.50	8.74
Total	Mean	4.6542	30.0375	54.4933	6.4637	7.4633
	Std. Error of Mean	.82521	1.05691	4.40850	.38842	.36898
	Minimum	1.60	23.70	33.90	3.66	4.50
	Maximum	16.00	50.20	109.20	9.13	10.40

Table 2: Mean, Std. Error, Minimum and Maximum influent and effluent sulfate, sulfide, TSS, BOD5 and COD concentration values of JWBO CW

code of the cell		Sulfate (SO42-), mg/L	Sulfide Ion (S2-), mg/L	Total Suspended Solid (TSS), mg/L	Biochemical Oxygen Demand (BOD), mg/L	Chemical Oxygen Demand (COD), mg/L
Influent	Mean	153.3333	4.5533	201.3333	273.3333	619.3333
	Std. Error of Mean	17.63834	.48015	6.69162	21.85813	187.25770
	Minimum	120.00	3.60	188.00	230.00	360.00
	Maximum	180.00	5.13	209.00	300.00	983.00
cell 1	Mean	39.6667	.0580	32.6667	6.3333	63.6667
	Std. Error of Mean	8.66667	.01856	11.40663	1.85592	14.49521
	Minimum	25.00	.04	15.00	4.00	36.00
	Maximum	55.00	.10	54.00	10.00	85.00
cell 2	Mean	26.3333	.0640	37.3333	7.3333	76.3333
	Std. Error of Mean	1.45297	.01604	10.17076	2.33333	17.64779
	Minimum	24.00	.05	21.00	5.00	47.00
	Maximum	29.00	.10	56.00	12.00	108.00
cell 3	Mean	41.0000	.0607	32.3333	5.3333	66.6667
	Std. Error of Mean	8.50490	.01676	12.03236	.88192	15.24613
	Minimum	28.00	.04	14.00	4.00	38.00
	Maximum	57.00	.09	55.00	7.00	90.00
cell 4	Mean	34.6667	.0557	33.0000	6.0000	63.3333
	Std. Error of Mean	1.45297	.02080	7.23418	2.64575	11.20020
	Minimum	32.00	.03	21.00	2.00	41.00
	Maximum	37.00	.10	46.00	11.00	76.00
cell 5	Mean	28.3333	.0593	34.0000	5.0000	80.3333
	Std. Error of Mean	3.17980	.01087	7.23418	.57735	20.53723
	Minimum	23.00	.05	22.00	4.00	49.00
	Maximum	34.00	.08	47.00	6.00	119.00
cell 6	Mean	35.6667	.0737	34.6667	4.3333	65.0000
	Std. Error of Mean	2.33333	.01317	7.51295	.88192	11.53256
	Minimum	32.00	.06	22.00	3.00	42.00
	Maximum	40.00	.10	48.00	6.00	78.00
Final effluent	Mean	34.3333	.0467	30.3333	2.0000	68.0000
	Std. Error of Mean	.88192	.01934	10.47749	.57735	11.71893
	Minimum	33.00	.02	17.00	1.00	46.00
	Maximum	36.00	.08	51.00	3.00	86.00
Total	Mean	49.1667	.6214	54.4583	38.7083	137.8333
	Std. Error of Mean	8.57631	.31393	11.90329	18.63877	42.89153
	Minimum	23.00	.02	14.00	1.00	36.00
	Maximum	180.00	5.13	209.00	300.00	983.00

Table 3: Mean, Std. Error, Minimum and Maximum influent and effluent TC, FC concentration Temperature and PH values of JWBO CW

code of the cell		Total Coliform (TC), CFU/100ml	Fecal coliform (FC), CFU/100ml	Wastewater Temperature, 0C	Wastewater pH, pH unit
Influent	Mean	60000000.00	4866666.6667	25.8000	7.1000
	Minimum	53000000.00	2600000.00	25.40	6.83
	Maximum	73000000.00	8500000.00	26.40	7.64
	Std. Error of Mean	6506407.099	1835150.614	.30551	.27000
cell 1	Mean	3500000.0000	480000.0000	23.7333	6.9767
	Minimum	3400000.00	470000.00	23.00	6.75
	Maximum	3600000.00	490000.00	24.70	7.10
	Std. Error of Mean	57735.02692	5773.50269	.50442	.11348
cell 2	Mean	3700000.0000	470000.0000	23.8333	6.9533
	Minimum	2800000.00	460000.00	23.80	6.58
	Maximum	4200000.00	480000.00	23.90	7.58
	Std. Error of Mean	450924.97528	5773.50269	.03333	.31524
cell 3	Mean	3566666.6667	490000.0000	23.3000	6.9700
	Minimum	2500000.00	480000.00	22.20	6.79
	Maximum	4700000.00	500000.00	23.90	7.08
	Std. Error of Mean	635959.46761	5773.50269	.55076	.09074
cell 4	Mean	3300000.0000	450000.0000	23.5333	7.2567
	Minimum	2300000.00	440000.00	22.30	7.22
	Maximum	4000000.00	460000.00	24.30	7.33
	Std. Error of Mean	513160.14394	5773.50269	.62272	.03667
cell 5	Mean	3733333.3333	440000.0000	23.5333	7.0067
	Minimum	3000000.00	430000.00	22.90	6.86
	Maximum	4400000.00	450000.00	24.10	7.09
	Std. Error of Mean	405517.50202	5773.50269	.34801	.07356
cell 6	Mean	3100000.0000	430000.0000	23.5667	7.0467
	Minimum	3000000.00	420000.00	22.30	7.00
	Maximum	3200000.00	440000.00	24.20	7.07
	Std. Error of Mean	57735.02692	5773.50269	.63333	.02333
Final effluent	Mean	3300000.0000	340000.0000	23.4000	7.1767
	Minimum	2000000.00	210000.00	22.60	7.15
	Maximum	5500000.00	560000.00	24.80	7.19
	Std. Error of Mean	1106044.002	110604.40015	.70238	.01333
Total	Mean	10525000.00	995833.3333	23.8375	7.0608
	Minimum	2000000.00	210000.00	22.20	6.58
	Maximum	73000000.00	8500000.00	26.40	7.64
	Std. Error of Mean	3961075.328	360399.79646	.21760	.05134

ANNEX II. ANOVA ANLYSES RESULTS

Table 1: ANOVA test results of orthophosphate and total phosphorus for the influent and effluent concentration values of each wetland cells of JWBO

```
Analysis of Variance
Source      DF        SS        MS        F          P
Factor       1       2.77      2.77      0.74      0.411
Error       10      37.71      3.77
Total       11      40.48
                                         Individual 95% CIs For Mean
                                         Based on Pooled StDev
Level        N      Mean      StDev   ------+---------+---------+---------+
Orthopho     6      6.915     1.898   (-----------*-----------)
Total Ph     6      7.877     1.985            (-----------*----------)
                                      ------+---------+---------+---------+
Pooled StDev =     1.942             6.0       7.5       9.0      10.5
```

Table 2: ANOVA test results of NO_3-N, NH_4^+-N, TN, PO_4-P, TP, Sulfate, Sulfide, TSS, BOD_5, and COD for wetland cell effluent concentration values versus wetland plant species.

```
Analysis of Variance
Source      DF        SS        MS         F          P
Factor       9    86080.5    9564.5    104.89      0.000
Error      170    15501.9      91.2
Total      179   101582.4
                                          Individual 95% CIs For Mean
                                          Based on Pooled StDev
Level        N      Mean      StDev   --+---------+---------+---------+----
Nitrate     18      2.894     0.761   (-*-)
Ammonium    18     28.978     2.608                (-*)
Total N     18     47.801     7.363                         (-*-)
PO₄³⁻P      18      6.313     1.935    (-*)
Total P     18      7.326     1.784    (-*-)
Sulfate     18     34.278     9.479                   (-*)
Sulfide     18      0.103     0.176   (-*-)
TSS         18      5.722     2.697    (*-)
BOD         18     69.222    23.511                              (-*)
COD         18     34.000    13.907                  (-*)
                                      --+---------+---------+---------+----
Pooled StDev =     9.549             0        25        50        75
```

ANNEX III: Pictures of Coliform Bacteria, Economical values of wetland plants and impacts of Wastewater on natural wetlands of CRV in Ethiopia.

1. Coliform Bacteria Pictures

Yellow Color TC colonies	Yellow color FC colonies

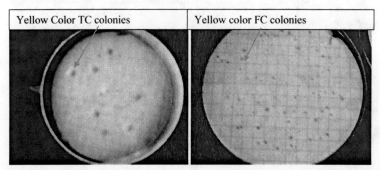

A. TC Colony Picture of cell 1 effluent   B. FC colony Picture of Cell 1 effluent

Figure 1: Indicator Bacteria; TC and FC colonies obtained at cell 1 effluent with 10^5 & 10^3 dilutions respectively.

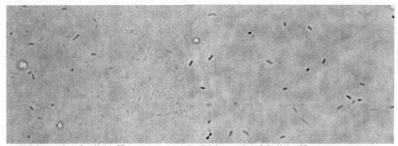

A. TC bacteria of cell 1 effluent   B. FC bacteria of Cell 1 effluent

Figure 2: TC and FC bacteria seen with Fluorescent Microscope

2. **Pictures which show some of the economical values of wetland plants in Ethiopia**

C. papyrus flower for selling at Bahir-Dar

C. papyrus for fishing at Bahir-Dar

Different materials made from *C. papyrus* Bahir-Dar

C. papyrus collected for use, Bahir-Dar

P. canariensis for city decoration, Bahir-Dar

3. **Picture of deferent wetland plants in natural wetlands of Ethiopia**

C. papyrus, Lake Tana, Bahir-Dar

P. canariensis, Finote-Selam, western Gojjam

C. alternifolia, Lake Awassa   C. alternifolia, lake Zeway   C. alternifolia, Deber-zeyet

4. **Pictures which show the impact of wastewater for natural wetland in CRV**

| Surface runoff from flower Farm, Lake Zeway | Tanney effluent, Chefee Meda, Tekure weha, Awassa | Flower farm effluent, lake Zeway |

DECLARATION

I, the undersigned, declare that this is my work and that all sources of materials used for this thesis have been duly acknowledged.

Birhanu Genet   Signature--------------------------   Date....................

This thesis has been submitted for examination with my approval as the research advisor of the candidate.

1. Dr. Seyuom Letta   ....
   University Supervisor   Signature   Date

CPSIA information can be obtained at www.ICGtesting.com
Printed in the USA
LVOW052041220812

295486LV00013B/46/P

9 783846 504390